The Englishman

Memoirs of a Psychobiologist

The Englishman
Memoirs of a Psychobiologist

by

John Staddon

The University of Buckingham Press

First published in Great Britain in 2016
by

The University of Buckingham Press
Yeomanry House
Hunter Street
Buckingham MK18 1EG

A CIP catalogue record for this book is available at the British Library

ISBN 978-1-908684-66-0

Printed and bound in Great Britain by
Marston Book Services Ltd, Oxfordshire

DEDICATION

To my Mother, Father and Grandmother for love, laughter and putting up with much nonsense.

Contents

Preface

When I retired from Duke University a few years ago I donated most of my files to the Archives of Psychology at the University of Akron. I had no idea that I might need them – to help me write a memoir, for example. But perhaps it was just as well. To include everything in a life would surely exhaust the writer, bore the reader and miss the whole point. Which is not to document every detail, but to try to show the pattern, to involve and perhaps even to entertain. A pudding may lack, as Winston Churchill once complained, a theme, but a life should have at least one or two.

Although I have been an academic for most of my life, the way I got there has taken some surprising turns. The first four chapters of this memoir describe what I can remember and discover about my early life: an unsuspected ancestry, fun in WW2 London, comical schooldays, and a spell in colonial Africa interrupting a wobbly college career at the end of which I left England for America. In the U.S. I followed again a slightly erratic graduate-school trajectory that ended up in a Harvard basement.

The main part of the book is about science, my efforts to understand the world opened up for me by biology, Darwin, the evolving cybernetic revolution and the experimental methods of influential and opinionated behaviorist B. F. Skinner. I have tried to make this part as simple and nontechnical as possible, although a couple of graphs have intruded.

My work has gone in several directions. First, study of the origins of learned behavior, behavioral *variation*. That part owes a debt both to Darwin, for the concepts, and Skinner for the experimental method. Second, study of animals' striking ability to tell time. How does it work? Is there a 'clock'? Does it involve – as I was to conclude – processes of *memory*? Third, the dynamics of reward schedules. Animals and people settle down to extraordinarily orderly patterns of behavior under such schedules. Laws can be derived. What do they mean? Most importantly: what do they tell us about the processes going on in individual organisms? (Surprisingly little, it turns out.) And finally, the similarities between behavioral ecology, learning psychology and economics. All try to understand their subject matter through the lens of rationality: does the species/pigeon/consumer act so as to maximize payoff? Maximization – rational behavior, Darwinian fitness – is a unifying idea, but in the end its failures are more scientifically interesting than its successes.

An academic career often involves travel. Although I never sought travel for its own sake, in fact I have spent much time working in more or less exotic, or at least remote, locations: Oxford (which can seem exotic), Australia, Germany, Italy, Mexico, Brazil, Argentina and Japan – and Canada. The next few chapters are about some of those experiences. The book ends with a dip into academic politics and some satirical reactions to it published in a now-notorious Duke University faculty newsletter.

This is not a tell-all biography. I have been married twice. My first marriage had its good parts – my two children, certainly, and mutual support at the stressful early stage of my career. But in the end the stresses proved too much. My second, although not without emergencies, has sustained two people happily through all our travels and travails. I owe my wife the greatest of debts for that. But few titillating details will be found in these pages. I hope that their absence will not disappoint too many readers.

Chapter 1: War Toddler

No child's home seems strange to him. Yet mine was much stranger than I knew. I was born in 1937 in Lavender Cottage, a small house in the village of Grayshott, on the Hampshire-Surrey border. The house had been rented by my mother's brother, Eric Rayner, when he came to England in around 1933 from Calcutta (now Kolkata), India, where he had been working as a journalist – for a publication with the wonderfully Kiplingesque name of *The Englishman.*

Eric came to England first. My mother, Dulce Norine, and widowed grandmother, Irene Florence Rayner, came later, from Rangoon, Burma, where they had been living in a flat at 97 Sandwith Road with Ram Saing, their longtime servant, and my grandmother's two beloved cocker spaniels, Toots and Brownie. In Rangoon, my mother had been secretary to the head of the Burmah Oil Company, Sir Kenneth Harper. In any event by the end of 1936, mother and grandmother were in Lavender Cottage. Very fortunately for them, as most of my mother's friends who remained in Burma died a few years later during the Japanese invasion. My father, Leonard John "Jack" Staddon, was away, in the army, still in Rangoon, but returned to England late in 1937.

Mother, father and grandmother moved to London after a year or so. First to Barons Court, Kingsbury, and then briefly to a flat in Wembley. In 1941 we moved to a rather drab, lower-middle-class area of north-west London called Cricklewood[1], a location then relatively cheap because quite inconvenient for getting into the center of the city. My mother had to take a bus and then the tube – underground – from Kilburn, to get to her job in Baker Street.

Once a village, Cricklewood's axis is the Edgware Road, which runs straight as a die NW from Marble Arch along the route of the old Roman road Watling Street. In the early 1940s, like its neighboring districts of Kilburn, Willesden, Harlesden, Neasden and Dollis Hill, Cricklewood consisted largely of street after street of late Victorian row houses with a few more spacious streets lined with detached and semi-detached houses. We wound up in the poorer side. The area was mostly Irish – pretty girls and rough boys. Cricklewood was in fact a kind of *entrepôt* for Irish immigrants. Their entertainment was provided by many pubs and the famous Galtymore Dance Club on Cricklewood Broadway. I remember Jimmy Shand and his band (Scottish, but hey, all Celts together!). I heard recordings when I learned to roller-skate there in the daytime. Their music seemed to jingle along in hypnotically repeating circles of sound. The

---

1 There is an excellent Wikipedia entry for Cricklewood:
http://en.wikipedia.org/wiki/Cricklewood

band members did manage to all stop at the same time, but I could never figure out how. Alas, the Galtymore closed in 2008, the current residents of Cricklewood having very different musical tastes. Check out Zadie Smith's novel *White Teeth* for the current multicultural state of play in Cricklewood.

My grandmother Irene Florence Rayner in Mandalay in 1919.

The middle class, mostly Jewish, lived in the larger houses to the South. All the houses had gardens, some quite spacious, which offered lawn-mowing work to young boys like me (with tiny barrel-type push mowers; power mowers were unheard of in those days). I found later that famed neurologist-writer Oliver Sacks grew up in a grand house on Mapesbury Road not far away from us, but I never met him – although I might have mowed his parents' lawn.

We were not Irish or Jewish. We were in fact very much the odd family out, although I was not aware of it at the time.

Originally, we – my mother, father, grandmother and, after 1942, little sister, Judy – just rented. First, briefly, in Heber Road, and then, our longtime home, the ground floor of a two-storey end-of-row house: 36 Cedar Road. It was within my family's modest means and the house had a good-size garden – always my Father's priority, I believe. The ground floor was just three rooms plus a tiny scullery kitchen, which served also for face-washing (and shaving when my father was on leave) for two or three adults and one, and then two, children.

The house was heated by an open coal fire in the middle room. The coal was delivered every week by a large cart drawn by a handsome dray

horse with 'blinkers' and fringed hooves. The blinkers were to prevent the horse being startled by goings on to his side. The coal was in large sacks carried through the house to a bunker at the back. The coal men wore leather caps with long flaps at the back to protect their clothes from the coal sacks. One man's cap had not a flap but a bag, which contained some kind of large growth – it was something a kid would notice and puzzle over. I never found out what it was. The unheated loo was part of the house, but had no interior access. It was reached by going outside, under a lean-to. The bath upstairs, with an ancient copper gas 'geyser' water heater, was shared with the upstairs tenants.

36 Cedar Road in the 1980s.

Our life in Cedar Road is vividly described in my sister Judy's lively memoir *Finding a Flame Lily*[2].

My father and mother met in Rangoon, where my father was a rifleman (private) and then lance-corporal. He served first in the King's Royal Rifle Corps (KRRs) from 1930 to 1937. Then for two years he was a civilian, and for a time worked as a male nurse in the small Manor House Hospital in nearby Golders Green (defunct since 2000). He re-enlisted in 1939 just before WW2 was declared. Finally, in 1944, thanks to his success in psychological tests, he transferred to the Intelligence Corps in India. Judging by a bunch of photos of warning signs that I found ("Think Twice Before Talking Once" and so forth), he worked on security.

---

[2] Judy Rawlinson *Finding a Flame Lily: A teenager in Africa.* Rawlinson Press, 2015.

My father's picture of a snake charmer in Rangoon, 1932.

Left: The only picture I have of my father's father, Henry Joseph Staddon, and his sister Eileen, in 1932. *Right*: My father's dance partner: Actress Merle Oberon.

My father and mother had very different histories. He grew up in East London in a highly dysfunctional family. Alcoholic mother and absent father, three sisters, and a brother who died in a motorcycle accident. His father, my paternal grandfather, was born in Plymouth. Staddon is a Plymouth name: check out Staddon Heights, Fort Staddon, Staddon Farm, Staddon Lane and many other Staddons in Plym. His eldest sister, Constance, emigrated to the U.S. when two daughters married GIs. We had no contact with my father's side of the family, largely because of my mother's disapproval of them. Was she right to shun them? I have no way to know.

My father left school at 14, worked until he was 16 then ran away and joined the army, lying about his age. He was hauled back by his mother – out of love for his wages rather than for him, I suspect. He had another go two years later. This time, he was successful. He was eventually posted to Burma and eventually to India where he spent much of the Second World War.

Dad was a good-looking lad, six years younger than my mother. They were both great dancers. He recollected with some pride dancing, somewhere in India, with beautiful young actress Merle Oberon, future star of (among many other films) *Wuthering Heights,* with Laurence Olivier and David Niven. Oberon claimed to have been born in Tasmania but was in fact Anglo-Indian…

Left: My father's mother Adeline Mary Staddon (neé Moore). **Right**: My father Leonard John "Jack" Staddon in India in 1932.

One of the 'security' signs my father must have been involved with during the war.

My father's attraction for my mother was obvious, despite his lowly career prospects. And she was 32, almost beyond marriageable age by the standards of the day.

By the time I was 2½, WW2 had begun. I have very few and feeble early memories, but one of them is being carried by someone down into the basement of our first Cricklewood flat in Heber Road (several streets in the neighborhood were named after Anglican bishops, Heber, Chichele and a couple of others) while sirens wailed and bombs and ack-ack guns detonated outside. This must have been late 1940 or early 1941, I think, during the blitz. I also remember flocks of barrage balloons, intended to deter low-flying planes, and searchlights illuminating the underside of clouds. But we were very lucky. During the war, all windows were striped with paper tape to limit flying glass. I suppose our windows must have been blown out once or twice. I remember one bomb site at the end of the road where the explosion must

Father Mother, JS and sister Judy in 1943.

have been powerful enough to break windows a hundred yards or more away. But I have no recollection of any close-by explosion until much later – 1944 and the period of "doodlebugs."

My Father was on active duty in England during the first part of the war – guarding the Queen Mother, Queen Mary, at Badminton, for a while. Badminton House is the grand home of the Duke of Beaufort, in Gloucestershire. Dad also spent time at Sturminster Newton, in Dorset and the Salisbury Plain and even guarding Hendon aerodrome, close to us. He took me and my cousin Benedick, a year or so older, to see and board a bomber on the airfield. I remember crying. I have always been puzzled at this

Left: **JS and Father early in the war.**
Right: **JS and Father in Las Palmas, on the way to Capetown, 1957.**

because I loved planes and machinery. But when I told my father many years later he said, "Oh no; it wasn't you who cried, it was Ben!" An odd false memory.

My memory is vague on exactly where and when my father was in England. But sometime during the war he was sent out to India – Karachi, I think – on the P & O liner *RMS Orion*, requisitioned as one of the more comfortable troop carriers. He remained in India, safe but remote from us, for the rest of the war. From satirist Michael Wharton's extraordinary autobiography, I learned that he also went to India on the *Orion*, probably on the very same long trip in convoy (via Cape Town in South Africa, as the Suez Canal was controlled by the Germans). Wharton sailed in the greatest comfort in the middle of the war – fancy meals, waiters and white linen and so forth. Amazingly despite his eccentricity and limited qualifications, he was a Lieutenant-Colonel. The 'other ranks' he says, traveled cattle-fashion, below decks. That was undoubtedly my father's fate.

The troops guarding Queen Mary (center) at Badminton. My father is, I think, 5ᵗʰ on the right of QM in the first row seated. (Signed: "Mary R, 1942")

Wharton – 'Peter Simple' – author for more than fifty years of the "Way of the World", a 4-day-a-week satire column in the London *Daily Telegraph* – is perhaps best known for the Prejudometer. This device is scientifically designed to solve the eternal problem of racial prejudice: who has it and how do we know? Peter Simple alerted his readers to a nifty solution:

THE Macpherson Report's[3] definition of a "racist incident" as "any incident perceived to be racist by the victim or any other person" is causing immense trouble and confusion for all concerned. Yet there is a simple answer. As I have pointed out before, the Racial Prejudometer was originally developed by the West Midland firm of Ethnicaids for use by the race relations industry, but is now available to everybody (ask your nearest race relations stockist).

Inexpensive and handy for pocket or handbag, you simply point it at any person (including yourself) you suspect of "racism", press the easy-

---

[3] Sir William Macpherson chaired a public inquiry into the murder of a young black man in England in 1999. This is an extract from the *Telegraph* column of April 13, 2001. Wharton's two-volume autobiography comprises *The Missing Will* and *A Dubious Codicil.*

to-find "action" button and read off the result in prejudons, the internationally recognised scientific unit of racial prejudice.

A satisfied client writes: "After reading the Macpherson Report, I began to worry about being racist. I was sleeping badly and losing my appetite. My job in an important call centre was at risk. My marriage was on the rocks.

RMS *Orion*, My father's WW2 troopship to India.

"Then a friend told me about the prejudometer. What a difference! As I began to use it regularly, all my worries about racism vanished! Now I sleep like a baby, eat like a horse and am so full of energy and keenness that I have been promoted call centre section leader. I have just returned from an idyllic 'second honeymoon' in Florida and feel like a million dollars. Thank you, Ethnicaids, for all you have done for me." (Name and address supplied).

This is only one of thousands of testimonials. Why, then, is the prejudometer not in use by everybody in Britain today? Is it because of an all too common fear of science and technology? This simple electronic device is admittedly not yet perfect. There have been incidents in London when black people, Indians, Pakistanis, Somalis, Chinese, Japanese and others have all been involved, causing their prejudometers to "over-read" and implode.

**V 1 *(top)* and early German jet plane in
the Deutsches Museum, Munich.**

"There are still some snags and headaches to be ironed out," says a spokesman for Ethnicaids. "But the backroom boys in our research division are working flat out, and one of these fine mornings they're going to come up with the complete answer. Then we'll all be able to think about racism not just some of the time but every minute of our lives."

It is surprising that despite the availability of a technical solution, racism still vexes much of the Western world.

As I grew up, with the blitz raging on London in 1941, we were 'evacuated' (as the word was) to Chisledon and then to Aldbourne in Wiltshire, along with many other London kids and usually, I believe, with my grandmother, 'Gran', who really raised us as my mother always worked. The intent of evacuation was to save children from the bombs. My mother was probably then in Northampton, following the re-location of Marks and Spencer's head office away from London.

I do remember time at a rather primitive farm, in Wiltshire, I think; I must have been about six. One of the owners had two Alsatian dogs a small black bitch and a larger, more conventionally colored male. I loved the dogs and spent hours with them. The little black one was especially tractable and I trained her in various ways – to the point that the lady who owned her evidently became quite jealous.

When I was six or seven we, Mother, Gran and me, visited an extraordinary lady called Dorothy Hartley, a friend of Eric and Daphne's.

Dorothy lived in a tiny Welsh village called Froncysyllte (we just called it *Vron*). She was a writer and social historian. Unimpeded by a college degree (but aided by some independent means), she co-wrote a six volume history: *Life and Work of the People of England* but became famous for her 1955 book, *Food in England.* My Mother used to tell anyone who would listen that during an early visit, my contribution was to take apart Dorothy's vacuum cleaner and, more surprisingly, actually put it together again. I do remember a much later visit to Vron. Dorothy still had her by now very elderly cat, who got in and out of the house through her own little door.

During this early period, I went to boarding schools at Fortis Green (in North London) and Hemel Hempstead, but I have only the vaguest recollection of either. I apparently got very sick – scarlet fever, I think – at one of them, but it left no memory trace.

Near the end of the war, in 1944, the Germans developed what amounts to a primitive, unguided drone, the *Vergeltungswaffe 1* or V1 "buzz bomb" – so called because it was powered by a buzzing pulse-jet motor. Some 9000 of these 'revenge weapons' were launched in 1944, most at London although the aim was not perfect. The rumor is that south London got it much worse than north because the Bletchley Park people somehow fed the German targeters false information about where the buzz-bombs landed. The idea was to get the Germans to send the V1s to the south, to the country and away from the docks and the center of London. This policy might now be controversial of course, like many such decisions – opening dams on the Mississippi in times of heavy rainfall so as to flood rural rather than urban areas for example. Or the classic ethical problem: do you switch the runaway railcar away from the group of children onto a line where only one man is working? I daresay these fine points did not deter wartime leaders from what they judged to be best for the war effort.

The warning air-raid siren is still evocative after many years, as I found when I heard it again for the first time during the ITV WW2 series *Danger UXB*, aired in the U.S. in the early '80s. One day, late in the war we heard the siren and all, myself, my sister and Gran, dived under what was called a Morrison shelter – basically a large metal table in your living room under which you could huddle for protection should the building be hit. But I could see out of the window. It was a doodlebug. I was just 7 years old but I knew three things about buzz-bombs: that first the light went out (when the motor ran out of fuel), then the buzzing ceased (sound traveling slower than light) and finally the bomb would crash and explode. I saw the V1 flying by, then saw the light (engine) go out, then the sound ceased and then, after a second or two, there was an enormous explosion which blew out all our windows. But we were spared. I was terrified of course. Smart enough to know that when the buzzing ceased the exploding was about to begin, but not smart enough to realize that if I could actually see the flame, the machine was probably heading away from me and would explode at some distance. In any event, no injuries, but a memorable experience.

The next day I walked down to gawp at the wreckage – this was a regular recreation for kids at that time. The V1 had destroyed a half dozen row houses on a neighboring street, Ivy Road, two on one side of the road, four on the other. The rebuilt area is now a small square with new buildings. I hope the current residents realize why their little neighborhood is so different from the rest of the street.

In among the wreckage of the blown-up buildings I saw a Morrison shelter. It was completely flattened by the collapsed bricks – not much protection there. I still remember my reaction to the sight. I also remember missing in succeeding days two rosy-cheeked brothers who used to walk by our house to school every day.

Top left: **Gran, Uncle Eric Rayner, Aunt Daphne and baby cousin Ben, 1936.** *Bottom* **Morrison shelter, issued to people in WW2 (named after the Labour minister Herbert M).**
Top right: **Our house and where the V1 landed in Ivy Road, as they are now.** *Below:* **Rayners and Staddons on the back stairs at 38 Well Walk, Christmas 1950.**

The shelter was named after Herbert Morrison, a wartime minister. There was another type, corrugated-iron Anderson shelters, that you could put up in your back garden. They were named after Sir John Anderson, the Lord Privy Seal, in charge of air-raid precautions before the war. Like the concrete shelter on the road right in front of our house, they were usually damp, cold and smelling of pee: nobody seemed to actually use them.

I remember little of the two war-ending days: VE day, May 8, 1945 and the less exciting (for us) VJ day a few months later. Surprisingly, I do remember the political atmosphere which led to the resounding defeat of Winston Churchill and the Tories and the installation of a Labour government with a huge majority at the first postwar general election. I got much of it from the radio: the BBC's Home Service and Light Programme. (There were only two stations: the Third Programme came later.) We knew the voices of the main BBC announcers – I particularly remember John Snagge – better than the voices of our own parents. Raymond Glendenning did the sport and everyone listened to live commentaries. My grandmother, all 4 ft 11 in of her, especially enjoyed boxing: "Hit 'im! Hit 'im!" she would cry if her favorite (Sugar Ray Robinson, for example) seemed lacking in initiative.

So I was well prepared for the overwhelming Labour victory in 1945. Clement Attlee, a bald little man with a mustache disturbingly similar to Adolf Hitler's, was the new Prime Minister – "A modest man, with much to be modest about" sneered Churchill with uncharacteristic meanness. The election result was a huge surprise to many outside the U.K. But it was no surprise to 8-year-old me, since everyone I knew said the same thing: "Churchill was good for the war, but Labour will be better for the peace." People remembered the pain of the 1930s and associated it with the Tories. Class hostility was stoked by the Labour people; and the country was ready for the huge socialist reforms promised by the 1942 Beveridge Report. The welfare state was swiftly implemented: free health care, nationalization of energy production and the railways, state pensions, free milk and meals in schools (I remember the milk, but I was indulged with home-made sandwiches so I avoided the school meals) – a utopian vision – all was in place within a year or two. Not good for GDP, of course; Britain was an economic disaster for several years after the war. Food rationing was finally abolished only in 1954 – four years after it had ended in Germany. I didn't really like candy – "sweets" – but I can remember running out to get some the day rationing ended. Despite their economic ill effects, the Labour policies put the U.K. on a course from which it deviated little in future decades.

And now back to how odd our family was. Gran was born in Calcutta, my mother in Thaton in Burma and then moved as an infant to Pa-an, a small community with only one other 'European' family. She was raised by Burmese *ayahs* and spoke Burmese by the time she was four, although she retained none of it in adulthood (or so she said). Her father, my grandfather, was born in Toungoo in Lower Burma. He was in the Burma Police. He was constantly shifted from one rural posting to another – a government policy designed to prevent the resurgence of dissident forces, apparently. He died at the age of just 39, in 1918 – of pernicious anemia, brought on by repeated attacks of malaria caught in remote sites with no medical help. My own father, a soldier for much of his life, was born in the East End and met my mother when he was posted to Rangoon in the '30s. All three of my parents spoke enough Hindustani, the lingua franca of the sub-continent, to exchange confidences in front of my sister and me.

My Mother on the left, Sir Kenneth Harper in the middle and the office staff of Burmah Oil, Rangoon, in the early 1930s.

That is all I really knew until just a few years ago when my father, after a tiff with my mother, spoke to me in his garden at Cove, near Farnborough. The conversation went roughly like this (you have to imagine the cockney accent):

Dad: You know, your mother is Anglo-Indian.
Me: ?!
Dad: Yes. When I wanted to marry her, I had to ask my commanding officer for permission. But he said "You can't marry her; she's Anglo-Indian."
Me: ?
Dad (proudly): Yeah, but I got her pregnant didn't I!

My parents' wedding day, Rangoon, August 24, 1936, my grandmother on the right – a modest affair, evidently!

Lavender Cottage, in 1937, with my mother and baby JS.

So apparently pregnancy trumped racism and I was the cause of their (mostly happy) marriage. But Dad had to leave the army at this point – which accounts for the two-year hiatus between 1937 and 1939, when he rejoined. Dad's return to England in 1937 also saved him from the horrors of the Japanese invasion of Burma which occurred a few years later.

I told this story just a few years ago to Phoebe who was my girlfriend during my last year in college. She responded "Oh, I always knew you were Anglo-Indian". I guess my short, dark, much-loved but slightly accented grandmother, who lived with me for my first two years in college, was a dead giveaway. Not to me though; her maiden name was Spanish (Mendieta), and so we attributed her dark complexion to her Spanish ancestry. Her accent, of course, seemed perfectly English to us. (And why did I not feel the sting of racial prejudice since my 'wog' origins were apparently obvious?)

My mother later worked briefly for a Miss (actually "Mrs", but those were the rules...) Orringe, who headed Marks and Spencer's shoe department at their head office. This did nothing to diminish the attentiveness to footwear she acquired from Gran.

When I checked my father's account with my mother a while later, she reluctantly agreed that, yes, she was Anglo-Indian, but immediately changed the subject. So I never learned the details. Nor can I reconcile his account with my mother's 1921 School Leaving Certificate, which lists her as 'European'. I did learn recently from my daughter Jessica that her "23 and me" DNA test assured her that she had "8% South Asian genes" (husband Mike rated zero). I guess that puts me at 16%.

My grandmother, in their flat at 97 Sandwith Road, Rangoon. On the back of this picture, my mother wrote:

Irene Florence Rayner – taken about 1934, when she was 51, in our flat in Rangoon – with Toots, one of her beloved cocker spaniels. Notice her elegant shoes – she always prided herself on well-fitting glacé kid shoes (she took a minute size 3) which she maintained was the hallmark of a well-dressed lady and infinitely more telling than silks and satins! I can still remember the scent of <u>real leather</u> when she took me into a shoe shop to be fitted out.

My mother never went on to college, although she did get an opportunity. Apparently she declined so as to look after my grandmother – not that she was really in need of help – and because brother Eric, a year or so older, could not go. Such was the deferential feminine ethos of those days.

During and after the war I went to several primary schools – St. Helens (which I remember going to by tricycle – although by the map the distance looks too great for that mode of transport at the age of 5 or 6) and, briefly, the local state primary: Mora Road (now just "Mora" – in English and Arabic), a ten minute walk from our house. I was not happy at Mora Road school. The kids were tough and had no difficulty recognizing how different I was. I was skinny, physically feeble and talked funny. Bullying was routine. So my parents, relatively poor as they were, looked for something better. The answer was Burgess Hill – suggested by my one-time communist/socialist uncle Eric Rayner. Before the war he used to sell Victor Gollancz's Left Book Club editions from a barrow in Camden Town, or so I was told. I still have a tattered and faded-red Left Book Club edition of George Orwell's famous *Road to Wigan Pier* from his stock.

Eric was by then probably no longer a socialist. I think it was in Calcutta that he first met the eccentric English journalist and broadcaster Malcolm Muggeridge, who later also worked on the *Daily Telegraph* and became a friend. Muggeridge converted from bohemian voluptuary to religious enthusiast in his later years. Eric's transition to conservatism was less dramatic. He had a good job as something-or-other night editor at the *Telegraph* and worked weekends at the BBC – and, I think, the *Sunday Times* – at various times. My aunt Daphne had a little independent money; she never had a regular job. She was a painter – my father would always refer rather uncharitably to a still-life of hers as "kippers on a plate". Dad had a cockney gift for mocking expression. He always referred to annoying insects as "flying aubreys", for example. I found out much later that these eponymous bugs were in fact named after Aubrey, a height-challenged fellow soldier in Karachi.

Uncle Eric worked nights all his professional life, a regimen that may have affected his health. He, Daphne and their three children Benedick, Annabel and Rosalind, lived in an elegant, rented 5-storey Georgian house, 38 Well Walk, in Hampstead with room bells to alert the servants in the basement (not that he had any), furniture from Heal's, an elegant little walled garden and a great view of central London and Saint Paul's cathedral from the top floor. The house next door has one of those blue plaques – this one for the painter John Constable. Eric used to walk his sweet Alsatian bitch, Shandy, just down the road on Hampstead Heath.

We always visited the Rayners on Christmas day. Their decorations were splendid: a huge tree covered with fancy baubles that reached the high ceiling of their living room – and many candles (this was before the days of electric tree lights and full-on "elf and safety"). We were definitely the poor relations. All this was to change after Dad went to Rhodesia and Eric retired.

I remember Eric saying that "your politics change with your income" – *verb. sap.* But he was a poor capitalist. He always rented, although the houses were usually quite elegant – his last London rental was in the wall of Greenwich Park (40 Maze Hill) with a private gate into the park and a view of the observatory. But when he retired he had little money to actually buy a house and wound up in an inexpensive area in Corby Glen near Grantham in Lincolnshire. The Well Walk house, owned for years by (I think) a charitable foundation, had to be sold when the owners were in financial difficulties. Houses on Well Walk now sell for sums not unadjacent (as the Brits say) to £10 million.

Hampstead was, and to some extent still is, a haven for upper-middle-class left-wingers[4]. Famous residents include Douglas Jay, a Labour minister and friend of Eric's – widely pilloried for saying that "In the case of nutrition and health, just as in the case of education, the gentleman in Whitehall really does know better what is good for people than the people

---

[4] Check out,http://www.rotaryinlondon.org/clubs/hampstead/?The_Area:Archive: Notable_Current_and_Former_Residents

know themselves", a trademark proclamation of technocratic arrogance. Other left-wing notables were Labour minister and writer Roy Jenkins, contraceptive pioneer Marie Stopes, communist scientist J. D. "Sage" Bernal, and Fabian Society icons Sydney and Beatrice Webb. Sigmund Freud lived there briefly and his daughter Anna's clinic is there still. The Soviet influence persists, with a twist, in the form of mega-rich Russian oligarchs who have bought very expensive properties around the Heath. One, Witanhurst by name, is reportedly just a tad smaller than Buckingham Palace – at least if you include its hangar-sized basement.

So I was to go to Burgess Hill. The school was in two rather shabby, elderly buildings in Oak Hill Park in Hampstead. They were demolished years ago and have since been replaced by expensive apartments one of which (I believe) contained for a while the tragicomic Peter Sellers. Its philosophy was to say the least eccentric – although not quite as eccentric as a 1961 Pathé documentary portrayal[5]. In 1957, at age 21, the school described itself as follows:

> The school can now hold fifty children. Including ten weekly boarders, and eleven adults. The children have complete freedom within the framework of a self-governing community. Our way of life is determined and regulated by the school meeting which takes place each week – and where for the past two years, decisions have been made by agreement. Thus there are no rules (imposed either by adults or by a majority vote)...

So a "progressive", basically anarchist, day school. It took girls and boys from 5 to 14. Kids could do whatever they wanted, more or less, which suited me fine. I recall that at St. Helens I had been coerced to do (but not to understand) long division. After two years at Burgess Hill, I had forgotten all. On the other hand, I was very happy at the school, learned a bit of metalwork, much enjoyed playing "cops and robbers" on nearby Hampstead Heath, and made a few friends. And kid freedom: every little boy had a penknife, and not one of those wussy little two-inchers either. Most had air rifles, some quite powerful. A friend called Fuest had a German Schmeisser ('as used by the Hitler youth'!) that was much admired. A few years later, my neighborhood friend Brian G. used to amuse himself throwing knives at trees in our back garden.

---

[5] http://www.britishpathe.com/video/beat-school-neg-version The Hampstead school was reconstituted in Hertfordshire in 1961, so this film is of a newer incarnation which must also have folded soon after.

My memory is of a totally loving childhood, even though the words "I love you" (always on the lips of every American family member these days) were never spoken by me or my parents. They would have seemed otiose and insincere. People in those days were much more 'behaviorist' than now: "it's not what you say, it's how you act" was the motto.

In a discussion of punishment with my mother many years later, I commented that I had never been spanked – my parents were very tolerant. "Oh no," my mother said, "We spanked you if you needed it!" I have no recollection of any such spanking. A trauma-induced 'repressed memory' no doubt! The extent of their tolerance was tested some years later, when I had become addicted to tropical fish. To keep them in breeding condition, my fish required regular feeding with fresh meat, in the form of bright red *Tubifex* worms, delivered wriggling on river mud. To keep these little buggers healthy until they became fish breakfast, constant running water was required. That was provided by the tap over our single kitchen sink, which was also used for washing and shaving. That my father put up with *Tubifex* is a tribute to his infinite good nature.

The world then was a much freer place, especially for children, in spite of (or perhaps because of?) the much more punitive discipline then common. Kids as young as seven or eight traveled by themselves all over the city by bus or bicycle. I was always interested in art and regularly visited the Tate Gallery (now Tate Britain) at an early age. At a time when there was little money for traveling exhibitions, most of its galleries were then filled with works it owned. I particularly admired some seven galleries of pictures by the 19th century English artist J. M. W. Turner – bequeathed by him posthumously to the nation. When I encountered the Impressionists a little later my first thought was "What's the big deal!" since Turner had anticipated their vivid representations of light and atmosphere more than fifty years earlier.

My family, like all but one or two others on our long street, had no car; although my mother and father both worked, we could not afford it. (The only car on our stretch belonged to a neighbor opposite: a tiny dark green Austin 7, which he polished obsessively every weekend.) So like every other 8-year old, I got to school on my own. I took an electric trolleybus (quiet, fast-to-accelerate and comfortable, but discontinued in the 1950s) and then another bus, from Cricklewood to Finchley Road, and then walked up Heath Drive to the school. This much time off-leash would probably now be illegal, at least in the U.S.[6] The buses also, since they were open at the back and you could hop on or off anytime they

---

[6] http://www.washingtonpost.com/local/education/maryland-couple-want-free-range-kids-but-not-all-do/2015/01/14/d406c0be-9c0f-11e4-bcfb-059ec7a93ddc_story.html?tid=pm_pop

stopped. Very handy, but it required a bus conductor to collect the fares, though he could do so while the bus was moving. So, more efficient for passengers and a 'job creator', but more costly for the bus company – in this case London Transport, i.e. the government

Burgess Hill was little concerned about safety, but I remember that they did draw the line and retrieve me after I climbed out of a window on to a second-floor (3rd floor U.S.) parapet. I crept around the building until I came to another window. Looking in I saw the teachers at a meeting – and of course, they saw me.

I recall just a few incidents from my time at the school. The year I left (so I must have been just 10) I went to a summer camp the school had organized on the Isle of Wight. As usual, we had freedom to do whatever we wanted. I had not, at that time, learned to swim – I think that occurred in the next year or so at the instigation of my mother at the now defunct[7] Finchley Road Baths. But I recall exploring on the beach and finding a float from a crashed seaplane. The war had ended only two years before and debris was common around the coast. The float was in effect a small boat and I recall paddling it around in the sea on my own (another violation of 'elf and safety).

The only other thing I remember from that camp was a certain sexual frisson associated with an older (i.e. 13 or 14!) girl called Nina. "Nina knew all about it" was the cry and her tent was the venue of choice for the older boys. A few years later, at a sort of kissing party for teenagers (things were more innocent then!), I remember a more powerful sexual epiphany. Kissing one girl in the half-light I suddenly had an overwhelming Eocene impulse to possess (in some way not yet fully defined) them *all*. I have not experienced this kind of polyamorous imperative quite so urgently since – just as well.

I did have a sort of girl friend at Burgess Hill. She was Jewish and wore glasses. She invited me to her parents' home in Golders Green to play tennis on their court. The house seemed opulent to me, especially the doors, which were beautiful natural wood. The doors in my house, like everything else, were painted ('Eau de Nil', was the prevailing color – a job lot of paint acquired by my father from work, I think).

I recall the names of two peers at Burgess Hill. One was Ernest Rodker. I thought he was older than me, but the internet says not. His father was one of the 'Whitechapel Boys', artists and writers who hung out at the Whitechapel Gallery. A woodworker, peace activist, gluer of door locks (in a hotel housing apartheid-tainted S. African rugby visitors) and a founding member of the Battersea Power Station Community Group, Ernest has evidently remained true to the spirit of the school. The

---

[7] Apparently the late-Victorian building was destroyed by fire in 1972.

Top left: **First volume of Martin Bernal's magnum opus.**
Center: **Martin Bernal.** ***Right***: **Martin's father "Sage"**
Bernal. ***Bottom:*** **Enoch Powell, Martin's bête noire.**

other name is Martin Bernal (Martin reminded me of a third, Stephen Mosbacher, many years later, but I remember little of him, beyond the fact that he was also a friend of Martin's). Martin B. was my best friend, but of course I lost touch with him when I moved to my last school, St. Marylebone Grammar, of which more in a moment. Martin moved on to progressive Dartington Hall in Devon, a boarding school for the "fee-paying intelligentsia", closed in 1987.

Martin who, sadly, died in 2013,[8] was the son of the famous physical chemist, communist and womanizer J.D. Bernal – known, without irony, as 'Sage'. JD was a brilliant scientist – and historian, his Marxist-inspired *Science in History* is still well worth reading.' Sage' was friendly guy: it was said of DNA pioneer Rosalind Franklin that she was the only female scientist within his orbit who had *not* slept with Desmond Bernal. JD was not, I learned later, married to Martin's mother, independently wealthy artist and patron Margaret Gardiner – a fact of which Martin apparently remained proud. She did often go by the title of Mrs. Bernal, however. But when queried, sometimes responded "There are many of us."

---

[8] http://www.nytimes.com/2013/06/23/arts/martin-bernal-black-athena-scholar-dies-at-76.html?_r=0

One should not be too censorious. J. D. Bernal's 'open-marriage' ideal was shared by his various amours and was a common one in the upper-class bohemia of those days (these days also, I see from a recent *Guardian* article[9]). Not that it always worked well. Bertrand Russell had a similar attitude, but I remember him being surprised and upset at his own jealousy when one of his wives followed the sexual party line. In any case, JD evidently remained on good terms with all his women.

Martin got his education in England, finishing at Cambridge. After a few years as a fellow of King's College, he left to teach Near Eastern studies at Cornell University in New York State for the rest of his life – a transatlantic history that parallels my own in some respects. He is best known for a series of provocative books, the first entitled *Black Athena: The Afroasiatic Roots of Classical Civilization: The Fabrication of Ancient Greece — 1785-1985*, which argued that the amazing civilization of classical Greece in fact had a Semitic and black-African origin. When Martin was featured in a national magazine many years later, with a picture, I realized that he was my childhood friend and wrote to him. We exchanged a couple of notes. In June, 1988, Martin complained that only right-wing journals had paid attention to his most recent book. He visited Duke University twice while I was teaching there, but I was unaware – even though a friend of my son Nick in fact entertained him on one occasion.

Martin was his father's son. J.D. Bernal never gave up his allegiance to Soviet-style communism. Martin's ideology was not so party-linked. But he clearly saw the world through a political lens. His mother apparently also had strong and sometime bizarre opinions. The reason Martin left Cambridge seems to be the offense he gave to many dons by inciting radical undergraduates to disrupt a formal dinner at Trinity College. The *casus belli* was the presence at the dinner (I don't think he was even a speaker) of Enoch Powell, a onetime Trinity fellow, and a brilliant, highly principled, but controversial, conservative, politician.

Powell was controversial because he had earlier pointed out rather vividly, in the famous 1968 "rivers of blood" speech (he was quoting the Roman poet Virgil), the potential dangers of unlimited numbers of foreign immigrants who did not share English culture. Many had no wish to assimilate and, as we see now, some of their British-born children are violently opposed to all things Western (see the Pakistani twins in Zadie Smith's prescient *White Teeth*). Powell's warning was sufficient for many, Martin included, to stigmatize him as a racist. Martin wished Powell treated as a pariah. He was apparently unwilling even to tolerate his

---

[9] http://www.theguardian.com/commentisfree/2015/dec/04/non-monogamy-showed-me-what-it-really-means-to-be-with-someone

presence in Cambridge, much less invite him for debate. When Martin failed to induce his colleagues to boycott the dinner, he incited radical students to picket and disrupt it. It was this, apparently, that burned his Cambridge bridge.

I'm afraid he followed the principle 'don't debate, destroy' from an early age. In his autobiography, he proudly recalls "tearing down Tory posters" on his walks to and from Burgess Hill at the age of 8 or 9. My communication with Martin ended when he sensed that we were no longer of the same mind on some issues.

Bernal *père* is a model example of what one might call 'the folly of the intellectual'. He never ceased to admire Josef Stalin, even after the purges and gulags were common knowledge. He was untroubled by the subversion of genetic science, and ruin of several gifted Russian geneticists, by Stalin's favorite, Trofim Lysenko, a latter-day Rasputin. JD eventually met Lysenko, and thought his 'science' was non-mathematical (correct), sort of like Darwin (!!). Despite his excellent *Science in History,* Bernal little understood qualitative, inductive science. Darwin reasoned from particulars to the general, from the many varied facts of biology and paleontology to the law of evolution by natural selection. Bernal was more Newtonian. He was brilliant at reasoning from the general laws of physics to the details of molecular structure. He could reconcile his admiration of Stalin with the evils of Stalinism because he *was* so clever. With no reflection at all, JD could always come up with a plausible justification that made Stalin a force for good – in the end. 'Eggs breaking? Omelet on the way!' His 'wife', Margaret Gardiner, equally opinionated but perhaps not so 'Jesuitical', not as clever – and not so easily able plausibly to invert the obvious – saw the truth about Stalin and was not impelled to rationalize it. Intellectuals are good at disguising or even flipping obvious facts. An intellectual in whom the faculty of logical reasoning is dominant, like 'Sage' Bernal, may be especially brilliant.

I never knew JD, but his cleverness in argument reminds me of G. K. Chesterton's aphorism about madmen: "If you argue with a madman, it is extremely probable that you will get the worst of it; for in many ways his mind moves all the quicker for not being delayed by the things that go with good judgment. He is not hampered by a sense of humour or by charity, or by the dumb certainties of experience. He is the more logical for losing certain sane affections. Indeed, the common phrase for insanity is in this respect a misleading one. The madman is not the man who has lost his reason. The madman is the man who has lost everything except his reason[10]." Too strong for JD, perhaps, but not inapposite.

---

[10] G. K. Chesterton *Orthodoxy*, The Maniac.

MY MOTHER'S RECOLLECTION

This is an extract from an autobiographical letter that my mother wrote to me in 1975.

The Secretariat, Rangoon

My grandfather was born in Seringapatam in the State of Mysore where his father's regiment was stationed and attended Army Schools in India. He joined Government service as a young man and retired as Superintendent of Telegraphs. On retirement he was given the job of Caretaker of the Secretariat in Rangoon: a vast red-brick building in the form of a square. With an inner quadrangle and beautifully kept lawns: our playground after school when we were kids. These are my earliest memories of Grandpa and the big house the family lived in – 104, Dalhousie Street: the old folks, Liz and Hugh upstairs, and Nellie and her family downstairs. They seemed comfortably off and had a carriage and a

Calcutta: Firpo's and the Zoffany in St. John's church, 1930s.

mare called Bess, which my father had given them – much to Mother's dismay because it belonged to her originally when Father had a pony-trap in the districts. At that time they employed a 'Chokra' (Hindustani for 'young boy') to train in the work of the house and kitchen and, years later when Mother and I had a flat, he came to us as cook-houseboy (Ram Saing). Grandpa had a stroke and had to stop working, after which he walked with a limp. He was at my Father's bedside just before he died and described him as the only one of his sons who had never spoken to him unkindly.

Grandpa was a stern but kindly man and – to use your phrase, John – 'pathologically' honest. [My uncle] Eric was the apple of his eye and he took a great interest in him when he stayed with the grandparents on first starting school. His wife was an invalid for years...but ruled the roost from her chair and supervised the preparation of meals, etc. I always found her forbidding and unlovable (although I never let on to my parents how I felt). I stayed with them for just one year of my schooling...and it was purgatory: I never really felt 'at home' unless my parents were there too.

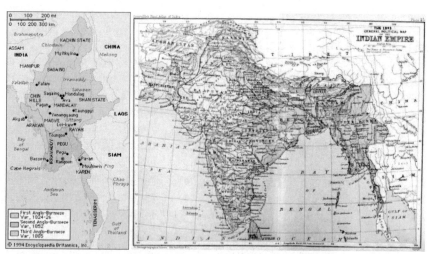

Left: **Colonial Burma. Thaton, my mother's birthplace, is just to the west of Pa-an, where she grew up as a child – north of Moulmein, in this map of nineteenth century Burma.**
Right: **The Indian Empire.**

Aunt Liz, the spinster living at home, was always kind and indulgent with us children, but she was at her office during the day. Uncle Hugh – who was no paragon of virtue himself – one day took the cane to Eric for some slight misdemeanor: I saw what was happening and rushed at him (quite uncharacteristically as I was usually docile and in awe of authority) to try and wrench the cane from his hand. I received a few flicks for my pains, but he did desist. There was a deuce of a row when my Father heard of Hugh's "presumption"....

Eric left Rangoon in 1924 (where he'd been a cub reporter/junior sub-editor with the Rangoon Times) to take up a post as Night Editor of *The Englishman*, at about three times the salary. Later, this newspaper merged with *The Statesman* (India's foremost daily) and Eric was Night Editor there for years. In the late 20s I visited Eric in Calcutta (Mother also made the trip later when he was married: he had a black cocker spaniel named Sally, and Mother brought back one of her pups, Toots, who with her daughter Brownie – remained with us till we left Burma), and he showed me around. I have a vivid recollection of walking across the "Maidan" (a huge, open space intersected by roads, in the more civilized part of the city) in the early morning "winter" mist, and thoroughly enjoying it. Among other places, we visited the Victoria Memorial (a Historical museum redolent of British/Indian encounters) and St. John's Church, which contained a painting of "The Last Supper" by Zoffany. Eric has always been deeply interested in history and used to walk miles exploring Calcutta in his quest for monuments and historical landmarks.

He also took me to the nightspot of the time Firpo's, and to the theatre to see the pantomime put on by the Calcutta Amateur Theatrical Society (the CATS) which was almost professional in its productions and had a great reputation. While he was in Rangoon he was able to take me to many shows (by virtue of the complimentary tickets he received as a newspaperman) and I can recall the unbelievable joy of seeing the incomparable Pavlova dance – the Swan, the Dancing Doll, etc. The Denishawn Dancers (the Americans Ruth St. Denis and Ted Shawn), the Co-optimists, No, No, Nanette and many others. A few years later, when I myself indulged in amateur theatricals (musical comedies), I was in a team of dancers who were once given brief lessons by Doris Humphreys (an American dancer who was touring with her Company). All very exciting to someone who adored music and dancing.

To revert to the family: my Mother's Mother, Florence Mendieta, soon married for the second time, one William Van Spall, whose father was one-time Governor of Dutch Cochin, a small enclave in S. India. He had three grown daughters by a previous marriage. My Grandmother quickly had a child by him, Ivy, and promptly lost interest in her daughter Irene [my grandmother]. The Van Spall progeny were always closer to her heart, and she doted on him. V.S. was a clever man and held a responsible post with the Government Secretariat in Calcutta. He was later sent to Rangoon to open up the Burma Secretariat. By nature he was odious and repugnant and I hated and feared him. Mother, too, despised him and slapped his face once for being rude to her Mother. Through drink and gambling, he squandered what little inheritance my Grandmother received from her first husband.

When he retired it was to my parents' home that he came (not to any of his own, indifferent daughters') and my father later fixed him up with a job and a house.

I was fond of my Grandmother (whom I always called Marmie) and I went to stay with her when I first started school at the age of 7½ – my parents being stationed at Magwe (near the Oilfields) at the time, where there was no school. Mother had taught me the rudiments of reading and writing, so I was in the Kindergarten for only 6 months before moving on to the dizzy heights of Standard I.

In all, I must have had 7 or 8 changes of school before I finished High School as a boarder in Maymyo (in the Shan Hills) just 2 months before my 17th birthday. I had caught up with Eric [just over a year older] by then and we both sat our finals together (he from the Govt. Boys' School in Maymyo and I from St. Michael's): our subjects differed in that he took History and Latin against my Geography and French. Our cousin, Clifford Scriven, also sat the exam (unlike us, he was excellent at Math and usually ended up with 100%) and was offered a scholarship to University as was Eric. My offer came later. Clifford's parents could afford to send him to England to train as a Chartered Accountant, but Eric and I had to refuse ours for lack of funds. [JS's mother later got an offer of college support from a Padre acquainted with the family, but turned it down for the reasons I have described.]

When her mother moved to Rangoon, my mother was sent as a "Foundationer" to the La Martinière School in Calcutta, which she left at the age of about 16 to join her Mother. There were two Martinière schools (boys' and girls') in Calcutta and two in Lucknow, founded by a Frenchman, Claude Martin, for orphans (originally), though they later took in paying pupils [evidently my grandmother counted as an orphan]. It had a good scholastic reputation in those days and discipline was very strict – e.g. cold 6 a.m. baths even in winter. Mother was always something of a rebel and was once put in 'solitary' on a diet of bread and water for speaking out of turn. She had a fine soprano voice and was in the School Cantata – they allowed her out to sing her part and returned her to her 'cell'! Mother loved recounting this story with amusement, and always spoke very highly of her headmistress, Miss Adams, who was a dignified Victorian lady.

My father [JS's grandfather], Henry Croft Rayner (Harry) was born in Toungoo in Lower Burma (Upper Burma had not yet been annexed) and had his schooling in the Rangoon College, where he finished High School and once won a prize for Latin: I remember seeing this around the house when I was a child. He was extremely good looking and I think Eric still has a photograph of him taken with his sister Nellie when he first joined the Police force, wearing his dark blue uniform and pillbox hat! Being fond of outdoor activities – swimming, riding, shooting – and there being few openings to choose from, I suppose he naturally gravitated towards the police, which he joined as a cadet. He was soon posted out into the districts and this proved his undoing, although he himself preferred the life to a desk job. He contracted malaria, which plagued him for the rest of his life, and he eventually died of pernicious anaemia, brought on by unrelenting bouts of malaria and sprue. He was constantly on sick-leave – on half-pay, of course – and could never afford to get away for a complete recuperation.

We were constantly on the move and lived camp-fashion in wooden Government bungalows. Away from the line of rail, Father had to do his touring by country boat and bullock cart (no Jeeps or Land Rovers those days) and he was often away from home for weeks at a time, leaving Mother alone with two small children. On one occasion, when he was chasing dacoits (armed bandits) he had to travel by elephant over the Siamese border south of Moulmein. If Mother wanted to get in touch with him she had to use a "runner", and this sometimes meant days. Astoundingly, she was never afraid to be alone and how she endured the life I can't imagine, because she was a sociable person and loved company. Her optimism and good spirits were fantastic in the face of the really hard times she went through. I can still see her sitting at her hand-machine, making clothes for us children, and singing her heart out. She was always warm and comforting and, although she didn't spare the rod, we always felt we were loved and safe with her around.

Both Eric and I were born in Thaton, on the Rangoon/Moulmein railway – Mother had Karen midwives for us (the women from this tribe made excellent nurses and are highly praised in the American Dr. Seagrave's book on the Burma campaign) and the doctor who attended her for Eric was drunk! She had a very hard time. We then moved to a remote village called Pa-an on the Salween River, a whole day's journey by launch from the nearest town, Moulmein. There were only two European families on this station – ourselves and the Robarts (they had two girls called Nancy and Geraldine whom we were friends with for some time). Mr Robarts was in the Excise Department and committed suicide by shooting himself: as there was no padre for miles, my father had to arrange the burial and read the service. I was very small then but can still remember our house up on the hill: my ayah used to take us for a walk every evening and we loved feeding the elephants at the bottom of the hill on their way home from work in the teak-yards. While in Pa-an, Eric contracted diphtheria and Mother had to rush him down to Moulmein by Government launch – reaching there just in time for him to be given the new serum which the hospital had only just acquired.

Father was then posted to Shwebo in dry Upper Burma (between the Irrawaddy and Chindwin Rivers) and his touring took him as far north as Myitkyina – this was blackwater fever country in those days and it did his health no good. Then to Magwe where I remember seeing Halley's Comet flashing across the night-sky (c. 1910), and where Father had an iron-grey pony called Jack whom I used to ride. Eric to school in Rangoon first and then me. Then to Hensada in the Irrawaddy Delta, and Prome, and Moulmein – where we were able to go to the church school. From here, Dad was posted to Amherst on the Coast, and we used to love the trips to the beach with our fox terrier to chase crabs in the warm, golden sand – with the lighthouse on Green Island flashing its light intermittently. Finally to Mandalay, where there was a newly-opened church school, and we were day scholars. Then Eric to Maymyo as a boarder where I joined him in 1919. In the meantime, Father was transferred to Shwegyin (on the Shwegyin River which was a tributary of the Salween), I was placed as a boarder in Mandalay and went home for the long summer holidays.

We travelled from the railhead in bullock carts in the dry season, and returned by cart, country boat and pony-cart in the rains when the monsoon had broken and the rivers were raging torrents.

Back to boarding school and then Father had his final and fatal illness – Mother had to bring him down to Rangoon through terrible conditions and he was extremely ill. And there he dies, but Mother was able to get us down from school to see him before the end. I had to sit my Middle School finals two months after this, but managed to get a scholarship which helped my Mother to send me to Maymyo as a boarder for the next two years of my schooling. The family – who were never very sympathetic – suggested that I should go out and work but Mother was insistent that I should go through high school. She only had a meagre insurance policy from my father but she took a post as a Matron of the school in Mandalay, and we were able to stay with her during our holidays.

On leaving school in early 1921, Eric went down to Rangoon to find himself a job, while I took a 3-months course in shorthand and typing. In May my Uncle got me a job with the Rangoon Port Commissioners, so down I went again, leaving Mother to join me the following year when she obtained a post as Matron of a Boy's School in Rangoon. During these years, she became a victim of dysentery and, when Eric and I were more on our feet, we persuaded her to give up working and we took a flat for the family in Rangoon. Some years later, and when Eric had moved to Calcutta, Phyllis Hurcomb [who remained a family friend in England later] joined us as a paying guest....

After six years' service with the Port Commissioners, where I was senior stenographer, I secured an excellent post with the Burmah Oil Company on 'covenanted' terms (i.e. on agreement with passage to England after each tour), as confidential stenographer to the General Manager. I was with them for 9 years and had a tidy sum accruing in my Provident Fund when I left the firm to get married and come to England.

My first leave to England was in 1933 and I can still remember the thrill of seeing Plymouth Hoe for the first time, as our P & O ship lay off anchor. What I had saved with BOC helped to pay the fares home of Mother and myself and to keep us going in Grayshott and in Kingsbury till your Dad came home from the army in October, 1937 – to see you for the first time...

Well, John, these are some glimpses of our very turbulent family life – at least in the early stages – and through all this I must pay tribute to the loving care and the fierce loyalty of my Mother – your Gran. Also to my dear Father who – despite his many faults and weaknesses – loved us and whom I loved and admired greatly. He was generous to a fault and a "soft touch" for any sponger who cared to take advantage of it – sometimes we the family suffered as a result, and he frequently caused my long-suffering Mother much distress and anxiety. How she ever kept the family together is a mystery to me – but then, women were cast in that mould in those days.

May their souls rest in peace.

Dulce Staddon
Cove, Farnborough, Hants.
29.10.75

Chapter 2: The Philological School

Burgess Hill left me in quite poor shape to succeed in the "Eleven Plus", the set of exams then taken by all children in their last year of primary school. These exams were a sort of IQ test, a product of the work of brilliant and subsequently controversial psychologist Sir Cyril Burt. Then they were a universal educational filter, based on three assumptions: that children differ in their intellectual capacities; that these capacities – IQ – are largely innate and more or less fixed by the age of eight; and that only the more intelligent can benefit from higher education. Pupils with parents rich enough to send to them to private school, a relatively small percentage of the total, were outside this system.

Based on the 11+ results, children were divided into two groups: one group went to so-called 'secondary-modern' schools, leaving at 16 and headed for a trade. The others, a small percentage, went to grammar schools, with two further hurdles to cross. First came Ordinary Levels, up to 10 three-hour exams in general subjects taken at the age of 16. If you did well enough in those you could go on for a further two or three years to the Advanced Level, in either "arts" or "science" – no more than four of those. If you passed A-levels you could go on to university. Only about 5% made it in 1955. There were few U.K. universities, perhaps a dozen or so, depending on what you count, and they cost the student nothing. No fees, and the local council would give you a modest bursary, just sufficient to live on.

Sir Cyril Burt, architect of the 11+ and much respected during his life, was attacked after his death. Oliver Gillie, a medical journalist for the London *Sunday Times,* reacting to a book by activist American psychologist Leon Kamin, pointed out some suspect data in a paper co-authored by Burt and two women. Numerical correlations from studies of identical twins, said to establish the high heritability of IQ, were exactly repeated in several ostensibly independent studies. Gillie also questioned the existence of Burt's co-authors, but after a while, evidence that they were not fictions came to light. The data were more troubling, even though the problems had been pointed out some years earlier by Arthur Jensen, an American individual-differences researcher and Burt supporter.

Was Burt dishonest? My guess is no. If you're going to fake data, why attract attention by inventing non-existent co-authors? And you certainly wouldn't repeat correlations exact to three decimal places. But Burt was probably sloppy. Like most other IQ researchers at that time, he believed that the high heritability of IQ was 'settled science', so he probably felt there was no need to be compulsive about data in line with a well-established conclusion. I think the vigor of the attack on him was motivated as much by a wish to undermine the hereditarian position as by genuine outrage at scientific malfeasance.

Top left: **"Raebourne" at Overton as it was. *Below:* Uncle Eric Rayner and Aunt Daphne, plus dog Monk, moving into their retirement house 5 Market Place, Corby Glen. *Top right: Tubifex* worms. *Bottom right:* An old picture of diatoms from one of Peter Quinn's slides, taken with Ivan's Rolleicord through my 19th century Watson microscope.**

Burt's best defense is his own undogmatic position on the nature-nurture issue:

[I]n these preliminary questions, it will be observed how modern psychology, adopting a biological standpoint, lays stress upon the genetic approach: it views the individual as the product of his past development. What special weight should be attached to each of the two main factors – the innate and the environmental – is a point about which current belief has constantly changed. From the time of Locke down to that of Bentham, English psychology tended to regard the new-born mind as a piece of unformed wax, moulded entirely by conditions acting after birth. With the inquiries of Darwin and Galton, opinion began to veer towards the opposite extreme. Some investigators even declared that 'nature is ten times as powerful as nurture'. Since the advent of the behaviourist and psycho-analyst, the pendulum has swung back once more; and heredity is held to be of smaller influence, at any rate in functional disorders. Biologists themselves would now probably agree that the psychologist's ideas of heredity have always been too crude and simple, and would be disposed to argue that any sure answer to such questions must await the conclusions of more exact research[1].

I did well on all the eleven-plus exams except mathematics – my abilities in that respect much impaired by the two fun years at Burgess Hill. My mother researched the options very thoroughly. I still have her copy of the 920-page *The Public and Preparatory Schools Year Book 1947,* published by the Dickensian firm of Gabbitas, Thring & Co. Ltd. For financial reasons, a real public (i.e. private school) was not for me, although apparently I did have a shot at the Haberdashers' Aske's School which was then nearby (they've since moved to Elstree in Hertfordshire). In the end it all came down to an interview with P. A. Wayne, somehow engineered by my mother. P. A. 'Pa" Wayne was headmaster of St. Marylebone Grammar School (SMGS) which, unlike Haberdashers', was not embraced by Messrs. Gabbitas, but was one of the schools to which I could have been admitted with a better eleven-plus score. Wayne had me read a passage from some Victorian writer. He asked me questions about it and about the meaning of various words. The only one I remember is "alms", which, at age 10, I must have got right. In any event, he admitted me to the school, which I began in the Autumn of 1947. (I found out many years later that I was not the only Burgess Hill pupil to make the transition to SMGS that year: Jack Watson was another – a big jump for both of us!)

SMGS, aka The Philological School, est. 1792, boasts among its old boys Marxist historian Eric Hobsbawm, writers Len Deighton and Patrick Russ (O'Brien) and William Willett, the inventor of daylight-saving time. It was a bit of shock after Burgess Hill. It was boys-only. Hence there was some physical bullying. The saddest victim was a boy called, I think Beurier, a small pretty, blond kid. We called him "Butter" and he was bullied a lot – though not by me, I think (not that I was a big help either).

All the teachers save one were male. The school was organized along 'public' lines, divided into four 'houses' with prefects and sub-prefects (6th-form boys with some disciplinary authority). I was in 'Moore House.' No 'fags' (prefects' servants) though. SMGS did the best it could, as a day school, to follow the public school model. We all wore uniforms and

---

[1] Burt, C. (1957) *The subnormal mind* 3rd edition. Oxford University Press, 1955, p. 13-14. The 'Burt controversy' is discussed at length by H. J. Eysenck in the Preface to the 1977 reprint.

Left: 38 Well Walk, Hampstead in 2015. And the Constable House plaque next door.
Right The old part of St. Marylebone Grammar School. The building is still used by Abercorn School for primary students. The ugly newer part, once the main school, has been replaced by an office building.

discipline was pretty strict.

I was way behind in math. My parents engaged the services of a 6th form boy, Frank Page, to tutor me in mathematics, which he did effectively enough that I was able to keep up if not excel (not that I was ever a very conscientious student!).

I remember a few incidents from my eight years at the school. When I was about twelve, I had my only fist fight, with another boy who was as skinny as I was. I don't remember the reason or his name and the result was inconclusive. I do remember struggling with math and the fearsome math teacher (to me, at least, but I think he was in fact really a rather kindly fellow) with the scary name of Mr. Snape[2]. I also remember – it must have been the 5th form, last year before O-levels – a history class with a new teacher. Our regular teacher, Dr. Derry, was not exciting but knew how to maintain order – he would occasionally bash an inattentive pupil on the head with a book. Mr. Hawke-Genn was the opposite. His lectures were interesting, but he was quite unable to control the class. I remember playing cards with my mates in the back of the room and marveling at our behavior: here was an interesting man yet, because he was unable to discipline us, we paid him no attention whatever. What a commentary on youth – male youth, at least!

The situation was otherwise with our art master "Arty" Leatham. He used to show us lantern slides, a few containing women in various states of déshabillé. Any disturbance and Arty would haul the offender to the front of the class, bend him over the desk, and whack him with a T-square. Discipline was good in Art and we actually learned something. I still remember a set of wonderful Dutch and Flemish flower paintings by Jan van Huysum and several others.

My performance in history was always poor.

---

[2] Yes, the very same name as the Harry Potter character Severus Snape, played by Alan Rickman. Was my Mr. Snape an inspiration for Ms. Rowling? I don't suppose so, but you never know.

SMGS in 1954. *Top:* **Pauliny (left arrow), JS (right arrow).**
Bottom: **Arrow***:* **Mike Hanson; P. A. Wayne in the center, Mr. Snape to the left. 3rd from right: Mr. Leatham, 4th Mr. Freudenberger and Mr. Finbow. Picture is 2/3 of total.**

Strangely perhaps, the teacher who influenced me most was a very dull German, Mr. Freudenberger ("Happy Mountaineer" – not a good name for him, really). Mr. F taught physics and he taught in a very rigid and, I suppose unimaginative way. I had difficulty with much of it – I particularly remember being puzzled then (and even now) by the concept of 'latent heat' as in 'latent heat of fusion' the heat given up when water freezes (go figure). The idea of 'internal resistance' (of a battery) also gave me a problem for a while. Yet I still retain from his teaching an appreciation for the beauty and order of classical physics. I found out later that Mr. F lived not too far from us, in an 8-storey block of flats called Ashford Court. He had a thick accent and was I suppose a refugee – Jewish? I don't know.

Then there was Mr. Hedges ("Duggie") an older teacher from whom I briefly took private piano lessons at his flat just up Lisson Grove from the school. He was a chain smoker, always lightly sprinkled with ash. I hated to read music and expected piano playing to come to me more easily than it did. Self-taught American jazz pianist Erroll Garner, who could not read music, was my model. I evidently lacked something possessed by Mr. Garner – *talent*, maybe. I'm sure I was a disappointment to Mr. Hedges – and, now, to myself. I wish I had stuck with it!

SMGS was on the corner of Lisson Grove and Marylebone Road. I remember (why does one remember trivia like this?) that the Lisson Grove bit was constructed of tarmac laid over wood blocks. As the tarmac wore, the wood was exposed. The wood was slippery when wet. Thick fogs were common in London in the 1940s because most homes were heated by open coal fires – they were outlawed in 1956. There was a 'great killer fog' in 1948. It was a source of quiet enjoyment for us young boys to watch in front of the school as cars, surprised by the stoplight looming through the fog, braked, then slid slowly but inexorably into one another. The fogs were fun for kids.

I went to Paris when I was at SMGS, on a school trip, for a week. We stayed in the Lycée Janson de Sailly, a French school in the 16th arrondissement, with a nice view but hole-in-the ground toilets that took some getting used to. I took a blurry box-Brownie picture of Napoleon's tomb and bought a small bottle of Veuve Clicquot champagne for my parents on the way back[3]. Thus, I acquired a solid basis in French culture.

Although I lack talent as a musician, I did love jazz and blues. In 1953, when I was 16, a terrible flood in East Anglia offered an opportunity to hear the best. Up until that time, the British musician's union blocked American musicians from performing in Britain. But then, in January 1953, a huge storm made a flood that killed more than 300 people in England and many more in the Low Countries. Norman Granz, the American impresario who had started Jazz at the Philharmonic in 1944, had the bright idea of offering to bring his musicians to England for a couple of 'flood fund' charity concerts. The unions couldn't really object to that and the concerts were held in the Gaumont State, a gigantic movie theater in nearby Kilburn, complete with Mighty Wurlitzer rising from the floor. I went to hear Ella Fitzgerald, Oscar Peterson, Zoot Sims, Gene Krupa, Herb Ellis, Lester Young and a handful of other jazz greats I had learned to love through their records. After the flood-fund concerts, American musicians began to work more regularly in Britain.

The JATP experience was not to be matched until 1978, when during a two week visit to the U.K. I took my son, Nick, and his cousins in my sister's Mini to Wembley Stadium to hear the Electric Light orchestra. I still remember the first note of Mik Kaminski's amplified violin shaking the huge hall from top to bottom.

I remember one other mega-entertainment from the 1950s. In 1954, with a local friend, Brian G, I cycled the few miles north to Harringay Arena, a large stadium normally devoted to greyhound racing. Our aim was to watch and be amused by American evangelist Billy Graham's London Crusade. We were late and crept into the vast stadium as

---

[3] Since I was but 12 years old, this purchase would now be illegal, I suppose.

inconspicuously as we could. We emerged in the middle of Graham's Thousand Voice Choir, which taught me convincingly the power of averaging. We stood next to a man who sang loudly but absolutely not in tune. Every note was off. Yet, the choir as whole was probably no worse than any other group of a thousand untrained singers. The only other thing I can remember from this event – which apparently was a very big hit at the time – is the preacher saying, in his broad American accent "and they hung the prophet Is<u>AY</u>ah [not Is<u>I</u>ah, as the Brits would say] from tree to tree…" gesturing broadly from one side of the great arena to the other. "Quite a stretch" we said. Alas, we were not brought to Christ, though many were.

I was never much interested in sport although when I was eight or nine I seem to remember that we played in the street a sort of cricket with a tennis ball. The pitch was across the street (no traffic then!) and the kerb was the crease. I remember two incidents at SMGS that prove my lack of sportiness. I could swim and was not too bad at backstroke. I was asked to join the swim team, a considerable honor to some, but declined. On another occasion, the sports master wanted to demonstrate the advantages of physical fitness. He chose two boys from the class: one was a top athlete, the other was me. We had to run up and down several flights of stairs. Then, Mr. Finbow measured our heart rates, hoping to show that the fitter boy's heart had accelerated less. The experiment failed. I did as well as the sportsman, possibly because I was as thin as a rail.

I did well enough in the O- ("ordinary") level exams, which I took at age 15, to go on to the 6th form and study for the A- ("advanced") levels, necessary for admission to a university. I had intentionally avoided taking O-level Latin, even though lack of it barred me from even applying to Oxford or Cambridge a couple of years later. Probably not a wise decision, but the irrationality of the requirement annoyed me – why should a scientist have to waste time learning Latin? (Why should the annoyance of a fifteen-year old have any bearing on his life choices?!)

There was another small hiccup. I had always been interested in science, so wanted to follow the science rather than the arts track (there were only two tracks). But P.A. Wayne, the forceful headmaster who had admitted me to the school, was rather disdainful of science and thought that my talents lay more in the humanities. So he put me into the arts track, which involved, among other things, learning German: which I did for a few months while petitioning to be shifted to the science track. I succeeded in moving to science, but of course missed several months of class in, for example, math.

Kids then took a maximum of four A-levels. I should have taken one in biology, because that is what I was really interested in. I used to walk from the bus stop in Edgware Road down Praed Street to the school and every day passed a little shop called Padbourne Aquariums. It contained many tanks of tropical fish. Mr. Padbourne believed in the 'balanced' aquarium, complete with plants and needing no artificial filtration or aeration. I was fascinated, having always been interested in bugs, butterflies and biology in general. I induced my very indulgent parents to let me have, at its peak, some six tropical aquaria (by then, we had the whole of 36 Cedar and so more room). My father got a friend of his who worked at the Handley-Page aeroplane works in Cricklewood to make me rustproof duralumin (an aluminum aircraft alloy) shades for my aquarium lights. Dad also insulated a small portable filing cabinet and added a handle so I could use it to transport tropical fish in two square bottles and keep them warm. My father was very adept at mechanical matters. If he had grown up in better circumstances he might have become a professional engineer. I learned many practical skills working with him in the little workshop he put together in our garden shed.

Continuing my biological kick, I bought an old Watson microscope (the model was "FRAM", I think, named after Fridjof Nansen's 1893 Arctic exploration ship) from Wallace Heaton and Co. in Bond Street. In those days, there was something called 'Resale Price Maintenance' which meant that shops everywhere, from Bond Street to High Street, were forced to charge the same for the same thing. This anti-competitive practice (abolished a few years later) meant that one tended to buy new items like watches from a high-end store – why not, the price was the same! So I shopped fairly regularly at Wallace Heaton's big store. The price for second hand items was not controlled, of course, but there were only a handful of places in London that dealt in old microscopes and WH was one of them.

I spent much time peering through my microscope, even taking some pictures with the aid of a friend's fancy camera (A Rolleicord; I also remember his self-winding Tissot watch, an expensive rarity in those days). I bought a book called *The Invertebrata* (1932, Borradaile, Potts, Eastham and Saunders were the authors) which had wonderful drawings of organisms from *Amoeba* to *Anopheles*. I collected British butterflies and also bought live stick insects and dead tropical butterflies which were sold in several shops around London. The shop I favored was in a little alley just off the Strand. I kept the butterflies in a nice cherry-wood case with several glass-topped drawers, for which I had paid 30 shillings. I had to give it all away when I left for America some years later. I also had to give away my Arthur Mee's *Children's Encyclopedia*[4], a wonderful collection of which I had read essentially everything but the fiction. I sold the microscope.

---

[4] Published by Grolier in the U.S. as *The Book of Knowledge*.

My camera-owning friend was Ivan Pauliny-Toth, a Czech, just arrived in England when his father, who had been Czech ambassador to Italy, fled the communist takeover in 1948. Ivan was my best friend. A year or so older than me, he was very smart. After only a year in the highly-selective SMGS, he was top in every subject, including English. There was then no shyness about comparing pupils' achievements. No one worried about damaging our self-esteem by showing the class just what everyone scored in every subject. Now, in most colleges, let alone high schools, students can't be publicly ranked even anonymously. Perhaps students now feel better about themselves. Whether they learn more than we did is less certain.

Ivan (I called him "Pauliny") and I used to spend time together on various informal science projects. We fiddled with government-surplus electronic junk, of which there was much right after the war – small cathode ray tubes, relays, capacitors and so on. This turned out to be very useful: much later, in graduate school at Harvard, I was amazed to find that my fellow first-year students did not know what an oscilloscope was. We also played chess. I always lost, until the very last game we played when I tried a new strategy. Instead of trying to look several moves ahead, I decided to look just at the next move, trying to find the one that left the board in the best condition. I remember that on this one occasion it worked, and I beat Ivan. No need for more games, then!

I fared less well in a school-wide essay contest. The theme was very general: I think that the topic was "Retrospection" or something like that. I wrote a ponderous piece on the ancient Egyptians, drawing heavily on the work of an American gentleman called J. H. Breasted (why do I still remember his name?). Maverick pharaoh Akhenaten and wife Nefertiti figured. Ivan wrote a witty parody of existentialism on the French theme "Les choses sont contre nous" which, I saw years later, may have been inspired by a 1948 Paul Jennings piece in *The Spectator* (or vice versa!). He won.

After SMGS, Ivan went to Cambridge where he was judged not good enough (!) to do nuclear physics for his PhD, but OK for radio astronomy, where he made a mark. He spent many years in the U.S. at the radio-astronomy observatory in Green Bank, West Virginia, then moved to the Max Planck Institute in Bonn (he spoke five languages so Germany was no problem). I saw him in Green Bank and once in Germany and then, finally in London, but after that we lost track of one another. He had a talent for satire, and many years later at my urging would contribute funny pieces about Germany to the Duke *Faculty Newsletter*, a curious and short-lived publication which I edited in the 1990s. (More about the *FN* later.) After two rather unhappy marriages and a history of intermittent ill-health, Ivan died sometime in the early 2000s, I think.

Green Bank was a strange place when my wife and I visited in the early 1960s. Our car (1952 Ford, bought for $200 from a graduate student and very unreliable) rounded a little mountain and approached the valley containing the radio telescopes, including a monstrous 300-ft one that collapsed in 1988. At the same time, the car-radio signal faded – the mountains blocked out most of the radio stations. The radio silence was presumably a boon for the radio astronomers, although, knowing how the political process works in the U.S., it was probably not the main reason the observatory was sited there. Ivan continued to live in isolated Green Bank even after the telescopes were controlled and data recorded remotely and most of his colleagues had moved to the urban delights of Charlottesville, home of the University of Virginia.

The height of my aquarium achievement during my grammar-school years came when, with another friend, I joined something called the Marble Arch Aquarist Society. I still have the lapel badge somewhere. I was about 15 at the time, I think. MAAS, a bunch of aquarium hobbyists, met once a week or so over a pub near the school. Smith (we called one another by surnames in those days) and I were the only youth in the club which had a couple of Fellows of the Royal Zoological Society as members. One, a Mr. Hervey, had written a book on the goldfish. My bible, however, was *Exotic Aquarium Fishes*, an American book privately published by the author William T. Innes. It has become a classic, going into 21 editions.

I spent a year as 'secretary' of the MAAS, writing the minutes of each meeting. I also did well in the quizzes, which seem to be something the Brits indulge in as part of any activity, from drinking (Pub quizzes) to watching tele – the BBC still has many prime-time quiz programs, some of them so opaque as to be almost incomprehensible. My memory is still cluttered with the Latin names of aquarium fish.

My A-level problem was that SMGS had no biology teacher until my last year there. So I never had a chance to study biology formally or take the appropriate A-level. Instead I had to take four other science subjects – Pure Math, Applied Math, Physics and Chemistry – only one of which I enjoyed much. I quite liked applied math and was adequate at it. For some reason I was always successful in chemistry, although I never found it very interesting and the teacher, Mr. Spinks, though popular, was not particularly exciting – at least not to me. However, the only prize I ever got at SMGS, which was shared, I think, was in chemistry. It was a book, which I could choose. I chose Bertrand Russell's *History of Western Philosophy*, rather an esoteric pick for a teenager studying science, I suppose (maybe Pa Wayne was right?).

I have been interested in philosophy from an early age. But I have always been rather 'literal'. I disbelieved in Santa Claus as soon as I compared the dimensions of our chimney with the size of a grown-up man. I have experienced few hints of spiritual feeling. I have never believed in a personal God. Henri Bergson and the *élan vital* struck me as mystical nonsense, although perhaps that was the influence of Russell, who in his *History* dismissed Bergson by saying that no one understood the *élan vital* except "ants, bees and Bergson". On the other hand, I don't think that religion is necessarily a terrible thing although some are, sometimes. I have sympathy for Gibbon's famous aphorism: "The various modes of worship, which prevailed in the Roman world, were all considered by the people, as equally true; by the philosopher, as equally false; and by the magistrate, as equally useful." The rise of monotheism has nixed the first, and atheistic fundamentalism the last. But all agree on Gibbon's middle claim. I guess Thomas Huxley's term 'agnostic' will do well enough for my religious views.

I was taught inorganic chemistry at SMGS. But I found organic chemistry much more interesting. I learned some when I was 14 or 15, during a family vacation at a cheap woodland holiday camp somewhere in the south of England. It was not a successful holiday: the weather was awful, it rained the whole time. Fortunately, I had taken with me a fascinating book: *Science for the Citizen* written by Lancelot Hogben and published in 1939. The book was intended to do just as its title says: teach science to the mass of the people. It is encyclopedic, more than 1100 pages of small type, and deals with every aspect of science from astronomy and atomic physics to the biology of behavior to eugenics. My copy of Hogben evidently came from the Marks and Spencer library – presumably a discard, since my mother would certainly have returned it otherwise.

Unlike most of his fellow left-wingers, Hogben opposed eugenics – state action to diminish the contribution of the less gifted to the next generation. The focus was on the poor and stupid, not so much on 'inferior' races, in the more homogeneous Britain of those days. Hogben thought eugenics a product of "the frustration which has accompanied the acceptance of sterility as the cardinal virtue of the middle classes. The eugenic movement has recruited its members from the childless rentier-twentieth-century bourbons who have earned nothing and begotten nothing." This is not the emollient language you would expect in a book intended to be popular. The 'bourbon' witticism would also elude most readers, I suspect. But Hogben may have a point about 'sterility', now that marriage and motherhood have become marginalized as career choices for middle-class women.

Despite its occasional deviations from the party line, SFTC has become a sort of classic. It is dedicated to Harold Laski, a charismatic and controversial socialist writer of the day. Ralph Miliband, father of the two brothers now active in Labour politics, was also an admirer of Laski. The book begins with a quotation from "A much abused writer": "up to the present philosophers have only interpreted the world, it is also necessary to change it". (That was Karl Marx).

Hogben was a fascinating character, in the general mold of Desmond Bernal (though not, as far as I know, a polyamorist). He was a committed communist, and dedicated to educating the masses – his other famous book is *Mathematics for the Million.* He proposed the skeleton of a universal language, *Interglossa,* early in his career and wanted to change the name of the left-wing Fabian Society to the Socialist Society. His zoological work on the Clawed Toad, *Xenopus,* was quite influential and yielded the 'Hogben Test' for pregnancy.

Looking back into *Science for the Citizen,* I see that even the chemistry chapter I absorbed over that rainy week has a socialist aura: it is called *The Last Resting Place of Spirits: A Planned Economy of Carbon Compounds.* The book is not an easy read because it pulls no punches. There is math! It presented the heart of each subject at quite an advanced level. I am impressed at my youthful self that I not only read the book but apparently enjoyed it.

SMGS was shut down by the Inner London Education Authority in 1981, despite an ineffective early intervention by Margaret Thatcher when she was education minister. Grammar schools had become a contentious issue in Britain. Many Labour ministers – Anthony Crosland was the most virulent – hated the idea of meritocracy or at least the inequality it implied. "If it's the last thing I do, I'm going to destroy every last fucking grammar school in England. And Wales. And Northern Ireland." Thus spake Mr. Crosland. The left wing wanted to destroy private schools as well, but in this they failed. Establishment forces were simply too strong and the government had no easy way to do it. The result was to eliminate a real upward path for talented lower-class kids, without compensating benefit. Because 'public' fee-paying schools, were untouched, eliminating almost all the grammar schools probably exaggerated social hierarchy, rather than reducing it. Grammar schools are back on the national agenda in 2015, apparently.

SMGS was all white. I have two whole-school pictures, from 1950 and 1954 – taken with a rotating camera with the 600 or so staff and pupils arrayed in a semi-circle – and there isn't a dark face to be seen in either. The reason was not racism but the homogeneity of the London population. Immigration from British colonies, in Africa and the Caribbean, only began in earnest in 1948, with the arrival in Tilbury docks of the *Empire Windrush,* from Kingston, Jamaica. The flow did not diminish in later years. London now, in 2015, is probably the most ethnically mixed major Western city, a huge change.

But there were many poor kids in grammar schools, at least at first. I confirmed this recently at a dinner of the *Old Philologians* – my first. I only discovered the OPs existence a couple of years ago. There were many of my generation. Very few spoke with 'posh' accents – indeed, one chap sounded just like my cockney father. My aquarist friend Smith at SMGS lived with his Scottish parents in just two rooms, with an outside water tap and toilet, in Paddington a fifteen-minute walk from the school (all buried now under the Westway Flyover). He and many other kids in the school came from very poor homes. With time, the mix probably changed as more middle and upper-class children passed the 11-plus –

Battersea Polytechnic, as it was.

perhaps with some coaching. A few not-so-bright middle class kids undoubtedly failed, to the great displeasure of their parents. So from both left and right, objections to the system began to be raised and the fate of the grammar schools was sealed.

In those days, your track to and in university was basically decided by the age of 16. A student was admitted not to the university (as in the U.S.) but to study in a particular department. I could not apply to do biology because I had not taken biology A-level. My record of A-level passes was not stellar, so I could not hope to get into the more prestigious fields, like physics, math or chemistry. The best I could hope for was acceptance by lesser subjects like medicine (yes, medicine was low on the prestige pole in those days) or engineering. In the end I was accepted to do chemical engineering at Battersea Polytechnic (now the University of Surrey, in Guildford), which was accredited to give University of London degrees.

It was not a good match. First-year students weren't even at the main campus in Battersea, but at the "Annexe" which was a converted shop in Putney, a district to the west. There our class, no more than fifteen or so, assembled every day from 9 to 5. In addition to lectures, the time was taken up with problem sets, breaking beams and similar mechanical exercises, and doing engineering drawings (no CAD in those days). It was a barrel of fun… On the plus side, I became friendly with a bunch of (slightly older) Greek students who hung out in a coffee bar in South Kensington. Alas, all I remember of that group is that my particular friend, Spyros, was always being ribbed about his birthplace: the island of Lesbos.

There were two girls in our class, one was Indian, the other Polish. I noticed that both of them completed problem sets much faster than I, which caused me to wonder if there was a subject in which I might not surpass them. Clearly, engineering did not qualify.

I had a friend from SMGS, Mike Hanson by name, who had gone to another college and knew more about the resources of the University of London of which, despite its "polytechnic" appellation, BP was a member. I discovered the University of London Union, ULU, in Bloomsbury near the ugly mini-skyscraper of the Senate house where my mother had worked, for the Ministry of Information, during the war. I noticed that there were many girls about, and a student newspaper, *Sennet*, to which I might contribute. I also heard of a subject called *psychology* about which I knew very little but which had many desirable features. For example, classes were not nine to five but perhaps four or five a week, and only the weekly tutorial was mandatory. You could skip the rest.

There were then also no A-level exams in psychology. How then were students admitted to do degrees in the subject? Why, by doing well on psychological tests! My only experience of such things was an IQ test which I took while I was at Burgess Hill, when I was around 8 or 9. I did pretty well – 150 or something like that. In any case, I scurried around and took the test that would allow me to transfer from Battersea to University College, London, for a BSc in psychology. I did well enough to be accepted so, again after a wasted term, I transferred. Once more I switched from a subject I did not like and was not good at to one I liked much better – although my reasons this time were not entirely intellectual.

Home life changed drastically the year I went to college. When the war ended my father, like most of his fellow soldiers, wanted out. But after he was demobbed he could get only a succession of unsatisfactory civilian jobs. Lacking either qualifications or contacts he began selling encyclopedias door-to-door. A soul-destroying job, but it came with a car – a tiny Ford Anglia.

In this chariot he covered the country staying in B & Bs and encountering temptations. He had a brief affair with one of his landladies, when I was about 14. I was of course extremely censorious. All the wrong was his, all the right my mother's and grandmother's. It didn't really occur to me that the presence of my grandmother, with both familial and chronological ties to my mother and to us – she was with us all during the war when my father was away, for example – a 'strong character', as my sister would put it, and no inclination (or even opportunity when space was limited) to give my parents much time alone – it didn't occur to me that my father actually had quite a beef. The matter was soon resolved but my father was in the doghouse for an indefinite time.

He went on to get a civil service job where he had a friend called Peter Quinn. Quinn was a bright, intellectually curious Irishman and he very kindly discarded in my direction a collection of old microscope slides. They included beautiful prepared slides of diatoms and foraminifera. All in nice wooden boxes, which I still have.

My mother, in the meantime continued to work for Marks and Spencer at their Baker Street head office (at the other end from Sherlock Holmes[5]!). She ended up working as the private secretary for the research director, Dr. Eric Kann. Hearing that I was interested in science, he passed on to her his copies of the weekly scientific journal *Nature*. The articles then were as obscure as they are now, but the possibilities they offered were intriguing.

A side benefit from working at the M & S head office was a cottage, "Raebourne", at Overton in Hampshire, in between Basingstoke and Andover. It was part of Berrydown, the estate of Harry Sacher, one of the M & S directors. The cottage was available for staff holidays, and some time early in the late 1940s we all went down there – the family by car, I by bicycle on a rather rainy day. A wet 60-mile ride along the A30 during which I consumed five breakfasts as fuel.

The cottage is still there, now extended. I remember that then near the cottage was a shed full of car batteries, which somehow provided electricity for the house. How they were charged, I know not. There were wonderful walks near Overton and I particularly remember the nearby trout River Test, with the fish – including a blind black one – swimming in the clear water under the little road bridge. All very familiar to country folk, but a revelation to us city dwellers!

---

[5] Holmes' address, 221, was in my day that of the Abbey National Building Society.

Finally, Dad got a quite tolerable job with William Moss, a local building firm, in charge of stores. With his practical and orderly nature, this was ideal. But his domestic situation rankled and in 1952, after balancing Australia and New Zealand with Africa, he emigrated on his own to Northern Rhodesia (now Zambia) to work for a private company. That lasted for a few months. But Dad saw that jobs with the colonial government were more secure and offered the promise of 6-month home leave every three years – an extraordinary holdover from the days when the colonies were a debilitating health risk. So, finally, he got a stores job with the Public Works Department. His first posting was to a small bush town called Kasama in the Northern Province. But his next was to an even smaller town with no more than 20 or so 'European' families. Fort Rosebery (now called Mansa, after the little river than runs through it) is in the Luapula Province. It was 150 miles by red laterite (dirt) road (often 'corrugated' by the passage of vehicles) across the 'pedicle' of the Belgian Congo, which reaches down into Rhodesia, from the nearest international airport in Ndola on the Copperbelt, the mining region from which N. Rhodesia derived most of its income.

My father on his way to N. Rhodesia via Capetown, in 1957.

When my father was securely settled, he offered my mother what amounted to an ultimatum: You and Judy must join me, or it's all over. Gran was not invited. My mother did join him, taking Judy who was still in school. So began the happiest years of their lives. But I remained, with Gran, to go to college and look after the house. Because of what was called the 'sitting tenant' rule then in effect, if a house owner wanted to sell, the current tenant got the best deal because he could not be evicted. Sometime in the early '50s we bought the whole of Cedar Road, which more than doubled our living space. So I had rooms to let.

When my parents left for Africa, my grandmother was 72 years old. She was in good health, more or less. She had only one eye, having lost the other to some kind of hemorrhage a few years earlier. But I, aged 18, also in good health and with both eyes, but surely lacking in maturity, was in charge – of the house, and of the various lodgers we had to help with living costs. The responsibility did not weigh as heavily on me as it probably should have. One lodger was a young secretarial girl but the rest were students, several of whom became good friends. Thus it was I entered on my first two college years.

Chapter 3: College Daze

The University College London (UCL) Psychology Department in 1955 was in two terrace houses (16 and 17, I think) Gordon Square, in the heart of Bloomsbury, on the other side of the square from number 46, the house where Virginia Woolf and her bohemian buddies lived in the interwar period. J. M. Keynes, most influential economist of the 20th Century, also lived there later.

University College, London, ca. 1950 and Jeremy Bentham and head.[1]

The first thing to be decided was my 'major' and 'minor' subjects. The terms don't mean the same as they do in the U.S. My real (U.S.-type) major was psychology, the U.K. major and minor were subordinate subjects. The major subject you took for two years, the minor you took for just one. Both were 'pass-fail', decided by a single exam at the end of the period.

What to choose? "Well, you were in engineering and you have two maths[2] A-levels. Perhaps maths should be your minor", so said my advisor. Bad choice. After all, I left engineering partly because I was *not* too good at maths. But, I had to take the advice and began to attend the lectures (a term late) and attempt the problem sets.

---

[1] The Jeremy Bentham 'auto-icon' with his mummified head at its feet. Bentham, founder of utilitarianism and spiritual 'father' of UCL, asked for his corpse to be dissected and mummified. The result, which you see here, is in a cupboard which "sits at the end of the South Cloisters on campus. He is usually woken up around 8am, and put to sleep at 6pm, Monday – Friday."

[2] U.K. "maths" = U.S. "math".

The textbook was brutal: G. H. Hardy's *Pure Mathematics*, a book as dry as the surface of the moon. Hardy, I later found out, was a wonderful man. He co-discovered the Hardy-Weinberg law, a mainstay of genetics. He alone responded to a letter from the poor Indian mathematical genius Srinivasa Ramanujan, and brought him to Cambridge. He wrote a splendid short autobiography, *A Mathematician's Apology*, which is still in print. But a great textbook writer he was not. And it didn't help that I had missed a term of classes. "The mathematical method had failed me", as a later influence, B. F. Skinner, might have said. In fact, *I* was failing and had to look around for something easier as a minor.

I picked social anthropology, a subject with a known high BS-quotient. The lecturer was a woman called Mary Douglas, who seemed to spend a lot of time talking about kinship relations in central African tribes. I did not find this interesting. The error was mine, of course. Mary Douglas went on to become professor of anthropology and a distinguished scholar. This was only the first of many such misjudgments on my part.

Like most university subjects in those days, pass or fail was decided by a single essay-type examination at the end of the year. Past exam papers were freely available. So I knew the sort of questions that would be asked. I also knew that the marking (grading) would be done anonymously by faculty chosen from across the several colleges of the University of London. They would not know who you were and you would not know them. In other words, I was perfectly free to study anthropology from sources other than Dr. Douglas. I chose a few fun books, such as Raymond Firth's *We the Tikopia* and Bronislaw Malinowski's works on the Trobriand Islanders, another Pacific island group with relaxed sexual habits. Both these gentlemen had a London association (though I did not know it): Malinowski was a Pole who became Professor of Anthropology at the London School of Economics and Firth succeeded him. I dipped into Margaret Mead's highly influential (but subsequently somewhat discredited) *Growing up in Samoa.* I had also developed a method of study. It involved lying on a sofa with a textbook, alternately reading and napping. No notes or exercises. But I suspect that the napping was crucial for memorizing. In any event, it worked for me at that time. So I was able to pass the exam at the end of my first year.

The story was much the same with my 'major' – economics. The text book was Stonier and Hague's *Textbook of Economic Theory*, published two years earlier. One of the authors was the lecturer – either Stonier or Hague, I cannot remember which. The book was not compelling and neither was the lecturer. So, again I followed the strategy which had worked so well for social anthropology. I even skipped the weekly econ tutorials after a while – a definite no-no, usually. Instead, I wandered the aisles of the giant bookstore Foyles on Charing Cross Road nearby, seeking a more readable alternative to S & H. I came across an American book by one Robert Samuelson: *Economics: An Introductory Analysis.* The book, then in its third edition, had been first published in 1948. It is currently in its 19th edition. It was to become one of the most successful texts of all time. In a word, it is a very good textbook. My clueless search for something better than Stonier and Hague had hit gold. So, no more classes.

But there was a problem when I turned up for the examination at the end of my second year. I was a day late! But, amazingly, the invigilator (that's what exam overseers were called then), suggested an 'out'. The University of London in those days offered two parallel degrees: *internal* and *external*. The syllabi were the same, but the exams and the exam takers were different. Internal degrees were for people enrolled full-time at one of the half-dozen colleges of the University of London. External degrees were taken usually by part-timers and students from the Empire (when there was one) – students at the University of Accra or Calcutta, say. But the exams were offered in London as well as at remote sites. So my invigilator, with a kindness well beyond the call of duty, suggested that I take instead of the internal exam which I had missed, the external one which was given at the same time the next day, i.e. today. This, I did. I passed, though whether this was because of weaker competition or because I actually knew the stuff I cannot say.

In the meantime, I much enjoyed life in college, not that I spent very much time there. I explored the University Union and discovered the weekly student newspaper, *Sennet* (subsequently *The London Student*) which was edited by a young woman a few years older than I. I don't recall her name, but she was possibly Jean Rook, who was to write regularly for the *Daily Express* and be hailed as the "First Lady/Bitch of Fleet Street" (versions differ) a few years later. I met a fellow student-journalist, Anne Berkeley, a kind girl and a gifted writer, who wrote entertaining weekly fictional episodes about a fresher female and her tribulations. I bought a Vespa motor scooter, a wonderful invention that made it possible to explore all of London (there weren't even parking meters in those days) at minimal cost and largely immune to traffic snarls. It also insulated me almost completely from the need to take any exercise, which was probably not a good thing.

The Vespa was a fun ride but not well-designed from a safety point of view. The wheels were small, only a third of the weight was on the front (steering) wheel and the engine was unsprung weight. All these features meant that the front wheel had only weak affection for the road. The vehicle first slipped from under me when travelling straight across an oil-slicked, cambered metal taxicab rank in the rain on Haverstock Hill in Hampstead. This was to become a habit in wet weather. I came off the

Top left: **Irene and a Young Conservatives 'ramble', ca. 1954.**
Top right: **125 cc 2-stroke Vespa and JS *en route* to the South**
of France. Note large, breakable windshield and various bits
of attached metal junk. *Bottom right* Kenny in Cassis, and
Irene on the balcony of our hotel in Diano Marina.

thing some eight times in two years. I only hurt myself the last time –
cracked tooth, mild concussion. I was carrying a friend inexperienced in
the ways of Vespas, so we fell awkwardly. (The tooth repair, by Mr.
Nevies, our Welsh dentist on Cricklewood Broadway, still holds after
nearly 60 years.) Few motorcycle riders wore crash helmets in those days.
Now they are required and the whole thing is much less fun. A crash
helmet would not have done me much good except, possibly, for my last
fall.

Before my father left for Africa, we had considered the possibility of
getting a car for me. It was an impossible idea, but looking was
entertaining. The car I remember was a 1938 Jaguar (this was in 1955 or
so, so the car was ancient). It was splendid-looking but much too
expensive (around £200, I think). Thank God I couldn't afford it;
maintaining it would have provided me with an unwanted career.

After introducing myself to the editor lady at *Sennet* I took the
initiative and visited Wardour Street in nearby Soho, where all the movie
companies then had their offices. "I represent 25,000[3] students of the
University of London, and we would like the opportunity to review your

---

[3] Things have changed: UCL alone now has 35,000 students.

films" was my pitch. In this way I got tickets to preview showings of a number of films. The showings were at 10:00 AM which was in fact a bit of problem, since I rarely rose before noon. This habit of late rising (no doubt a medical/psychological problem – not my fault, surely?) also interfered with my attendance at lectures. Most of the movies were from the U.S.; I remember particularly *Moby Dick*, starring among others Gregory Peck and Orson Welles, and *War and Peace* (Audrey Hepburn, Mel Ferrer). I wrote reviews of these movies. The reviews were not distinguished. I wanted them to be funny and that was much easier if the movie was bad. Unfortunately, these two were pretty good; the reviews, not so much.

I do remember one seriously bad movie, *Apache Ambush,* which I saw with a bunch of friends in 1957 or so. It starred no one of note and I can't remember exactly why it was so awful – or why we went at all. But awful it was. The fact that I remember its valence, but little else, proved to be a general principle. The last thing you forget about something is whether or not you liked it.

My evaluation of *Apache Ambush* was colored by the fact that we saw it in a theater in Leicester Square that had just acquired a new feature called "Dancing Water" or something like that, a set of fountains that rose up in front of the screen and spurted and swayed – and hissed – in time to recorded music. This spectacle by itself was hilarious. The image of massed schoolboys all prone and pissing in synch was irresistible.

In addition to movies, I also reviewed art shows. I have little memory of details, but I do recall an empirical finding that we called 'Bowenizing' after a show in Denis Bowen's New Vision Centre Gallery near Marble Arch. The show combined obscure modern art with long jargon-filled descriptions. It led me to formulate a law which has application well beyond New Vision: the quality of a piece of art is inversely related to the length of the accompanying description. "Staddon's Law" has few exceptions.

Not that I was averse to modern art. The one review of mine I can find spoke highly of two of Bowen's exhibitors: Albert Berbank and Otway McCannell. Their work still looks pretty good to me. I bought many art books and enjoyed even such extremities as Kandinsky and Malevich, artists I now find it very hard to appreciate. I liked Mondrian, though I preferred, and still prefer, the similar style of Ben Nicholson, husband of more-famous sculptress Barbara Hepworth (I found out much later that both had Margaret Gardiner, of whom I have already spoken, as patroness). I enjoyed Salvador Dali's *Corpus Hypercubus,* a sketch of which I used to illustrate a review. I still have a little original etching by futurist C. R. W. Nevinson whose WW1 *La Mitrailleuse* I had seen in the Tate. I bought a signed copy of a limited-edition *The Apes of God*, a book written and illustrated by vorticist Wyndham Lewis. I liked his art but didn't realize what a suspect fellow he was politically. It was unwise, even in 1931, to consider Adolf Hitler a "man of peace" – although Lewis was far from alone at that time.

I perpetrated some art myself, ink drawings mainly, which now look very much of their time. One such product, tragically lost for ever to the world of art (although a sketch remains), was a mural depicting iconic jazz tenorman Lester Young with his distinctive hat. I did it for the 2i's coffee bar in Old Compton Street, in which cockney rocker Tommy Steele, as yet not famous, used to play. The location now sports one of those commemorative medallions as "The birthplace of British Rock and Roll..."

'Bowenizing' anticipated what was to become the mode in modern art.

In 1974 Tom Wolfe could write in his little book *The Painted Word* "Modern Art has become completely literary: the paintings and other works exist only to illustrate the text." The U.S. pioneer was apparently *New York Times* art critic Hilton Kramer who complained that "Realism...does rather conspicuously lack a persuasive theory" to which critics might have responded "So what!" Unfortunately, none did. Kramer is an unfortunate target for criticism in some respects since he was a great defender of art as an aesthetic experience – rather than a political act or a form of self-medication. He was not a fan of the great nonsense of 'conceptual art' and the cartoonishness of 'pop'. But he did 'Bowenize' from time to time.

In a letter to me many years later, Wolfe recounted a related incident:

> I was once at a forum on the arts at which the architect Richard Meier gave a slide presentation of some of his new European work. He pointed to a slide of one of his stark white buildings amidst buildings that went back several centuries in some cases, and he commented: "You will notice that this structure sets up a dialogue with the

surrounding built environment." The late Harvard philosopher, Robert Nozick, raised his hand and asked Meier, "What did the surrounding built environment say?" Meier stared at him for what seemed like a full minute, without a word, then clicked that slide off the screen and went on to something else.

Despite a few checks like this, pretentious nonsense still abounds in art criticism.

Robert Nozick was an interesting philosopher who died too young in 2002 at the age of 63. In his *Anarchy, State, and Utopia* he pointed out

flaws in his colleague John Rawls' hugely influential *A Theory of Justice*. In particular, he criticized Rawls' assumptions that society should be arranged so that any change benefits the poorest most, and that not knowing their future position (Rawls' 'veil of ignorance'), all will prefer a society in which everyone is equal. Neither assumption is axiomatic. Many people prefer a society in which there is some chance of advancement. Not everyone thinks that every poor person should be helped before any richer person. Since Rawls' two core assumptions are not universally accepted, it is hard to understand why he continues to eclipse Nozick as a moral philosopher.

My only other foray into the world of art was when I later took my friend Phoebe to interview Slade sculptor Reg Butler after he won an international competition to depict the 'Unknown Political Prisoner'. I was impressed with the simple elegance of his house in Berkhamsted. Phoebe hung out with drama types like Roy Battersby and Tom Courtenay, later famous for roles in *The Loneliness of the Long-Distance Runner, Billy Liar*, and many others. Battersby was involved with many TV productions, but was apparently blacklisted by the BBC for his activities in the Workers Revolutionary Party, a Trotskyist group. Why the BBC, not noted for its hostility to left-wing ideas, found Battersby's actions disagreeable is hard to fathom.

Henry Forti, now become actor Henry Burn Forti, was a memorable hanger on. I don't think he was an official student; he was younger than me, but oh so much more sophisticated. In memory, he reminds me very much of the wonderful Anthony Blanche character in the Granada TV's iconic dramatization of Evelyn Waugh's *Brideshead Revisited*.

Another notable was young Gavrik Losey, a good-looking, fine-featured lad, the son of American movie director Joe Losey. Joe was blacklisted during the McCarthy years because of his communist associations. Although McCarthy and Congress get the blame for attacks on people like Losey, it was in fact the private sector – movie company owners and theater directors – who did much of the dirty work. Reclusive oddball Howard Hughes, owner of RKO Pictures, was apparently the first one to make Losey's life difficult enough that he was forced to emigrate. After failing to get work at RKO, Losey fled to England, where he directed a number of memorable movies, such as *Modesty Blaise, The Servant* and *Accident*. Son Gavrik also went on to a movie career.

Top left **Picasso-ish sketch by JS, ca. 1956.** *Top right* **Patricia, on the steps of UCL, circa 1957.** *Bottom right* **Phoebe in 1960.**

Political correctness is the new McCarthyism. All too often, some verbal misstep has proved devastating to the likes of Tim Hunt and Larry Summers. Even a thought-crime will be punished, as in the case of Los Angeles Clippers owner Donald Sterling who was clobbered for a private conversation. And again, their lives have been blighted not by the government but by institutions like universities or the NBA. Private enterprise may be free; it is rarely courageous.

We socialized in coffee bars but ate, usually, in Cypriot restaurants in a little street now destroyed by the erection of the egregious 33-storey

Centre Point on the same site, at the intersection of Charing Cross Road and Oxford Street. Obliteration of the past is now being completed, by a huge Crossrail excavation at that same intersection. Centre Point was an early victory in the war between skyscrapers and London's height restrictions. But it was vacant for many years before taste and tradition were finally defeated with the erection of the gherkin, Canary wharf and most recently 'The Shard'. Centre Point is too tall, of course, and destroyed a human-sized neighborhood. But it is less ugly than many that came after.

UCL, like most other U.K. universities, now emulates U.S. practice in its fund-raising efforts. It is rather effective. As part of this endeavor, a

few years ago it sent me a copy of *Pi*, the college newspaper. The date was February 16, 1960. How clever! I thought, to send me a paper from the year in which I graduated. But the gift was even cleverer than that for, in the center of the paper was an article on the 'Beat Generation': "Society has betrayed us, so we owe it nothing" the headline proclaimed. The article was illustrated by yours truly, something I had completely forgotten (I was not responsible for the content!). They got a good donation that year.

I was very much less impressed in 2015 when UCL President Michael Arthur, bowled over by a 'tweetstorm', urged and then accepted the resignation of Honorary Professor Tim Hunt, a Nobelist and emeritus (i.e. elderly) professor. Hunt was guilty of a couple of sentences of flippant and mildly sexist comments about women in the lab during a speech in Korea: "You fall in love with them, they fall in love with you and when you criticize them, they cry." President Arthur acted apparently without even talking to Hunt. Many distinguished colleagues all over the world subsequently protested, but the deed was done, and much applauded by the ritually sensitive.

The best response I've seen to Hunt's misstep was a bunch of funny tweets by female scientists doing macho things like exploring the arctic or wearing hazmat outfits. President Arthur should look and learn.

The Hunt incident reminded me of an earlier occasion, when a university president was the victim rather than the source of bad behavior. In 2005 Harvard President Lawrence Summers made some comments about women in science. Summers is not the most charming of men. Duke colleagues who have been to Harvard to sit on doctoral committees found him gruff and dismissive to the point of rudeness. He had already upset the Harvard faculty in various ways. But what he said should not have caused the storm it did. *The Economist* – not an illiberal source – in a piece entitled *Harvard's Disgrace*[4] reported the incident this way:

> In a talk at the National Bureau of Economic Research in January, [Summers] referred to three theories that might explain why women are under-represented in the highest reaches of maths and science: that they face discrimination and other forms of social pressure; that careers in so demanding a realm, calling for an 80-hours-a-week commitment, are less appealing on average to women than to men; and that the variability of aptitudes for science and maths differs between the sexes (leading you to expect that more men than women would be very bad at science, and more very good, regardless of the respective averages, a possibility for which there appears to be some evidence). Admitting to be no expert in the field, he then speculated, as he had been invited to, about which factors matter most. He said he suspected that discrimination was relatively unimportant, and that the variability of aptitudes might matter a lot.

These comments – factual followed by an opinion he was asked to express – were the final straw that caused Summers soon to resign from the Harvard presidency. But – compounding the problem – not before repeatedly prostrating himself before his accusers.

One of Summers' strongest critics managed to exemplify the very stereotype for which Hunt was excoriated ten years later. This is what she said: "Nancy Hopkins, a biologist at Massachusetts Institute of Technology [and member of the National Academy of Sciences], walked out on Summers' talk, saying later that if she hadn't left, 'felt I was going to be sick,' that 'my heart was pounding and my breath was shallow,' that 'I just couldn't breathe, because this kind of bias makes me physically ill,' and that she had to flee the room because otherwise 'I would've either blacked out or thrown up.'" AKA 'a fit of the vapors'? But no impediment to career success, apparently. Since I am a big fan of zebrafish and have done research on them myself, it saddens me greatly to see that Dr. Hopkins, a zebrafish expert, should have so let down our shared species. I wrote a little parody of the Summers incident (at the end of this chapter) but I'm afraid it's rather eclipsed by the reality.

As the Hunt incident shows, the hypersensitivity of universities in the both the U.K. and the U.S. to the feelings of any group perceived as 'disadvantaged' – no matter how ridiculous they may be and no matter

---

[4] March 17th 2005.

how much they conflict with academic freedom and the first amendment – has only gotten worse over the years.

The teachers in psychology at UCL were generally good, much more interesting than the Putney engineers. I particularly remember three: Henry James, who taught learning and perception, my tutor R. T. "Bob" Green, and A. R. Jonckheere, a brilliant but eccentric statistician. James was an excellent lecturer who wound up at Dalhousie University in Canada where he successfully built a new department. But his lack of respect for the 'softer' parts of psychology evidently got him into some trouble when he was promoted to Dean.

In addition to teaching at UCL, Bob Green also ran some kind of market research company. He was both easy-going and engaging. I do remember him demonstrating hypnotism during a Psychology Department get-together in historic Beatrice Webb House, in Holmbury St. Mary near Dorking. I tried the method out successfully with a couple of friends. But one insisted later that she was "just pretending". I never attempted it again.

Bob later got into some kind of battle with Marxist neurobiologist Steven Rose, a champion – along with Richard Lewontin of Harvard and psychologist Leon Kamin – of the 'radical science movement'. The fight was over Bob's possible appointment as the first Professor of Psychology at the Open University, a creative new higher-education enterprise launched by the government in 1969. Bob lost. I don't know the details, but I daresay Bob's activity in a private business was not viewed favorably. Rose, who was the OU's first academic appointment, was a relentless polemical infighter, passionately opposed to evolutionary psychology – possibly to any biological psychology at all.

A. R. Jonckheere (1920-2005) – "Jonck" – taught statistics and made significant contributions to the field. He was much loved, and the last UCL faculty of my era I saw – on a visit many years after I left, when the department had moved to a horrible 1960s building nearby. He frequently used a pre-power-point device called an overhead projector, a machine which lit a book, say, with a very bright light so the page could be projected on to a screen. I can still see him leaning over the projector, wreathed in cigarette smoke caught in the beam, as he explained some arcane point.

There was no animal psychology research at UCL. I learned about it for my exams, though, mainly through a massive American textbook: Osgood's *Method and Theory in Experimental Psychology* (1953). I tentatively inquired if there was any way I could do an experimental project with protozoa, combining my interest in biology with expertise in microscopy. The answer was no. The experiments we did do or (in my case, sometimes) pretended to do were mostly Mickey Mouse. I carried out a number of IQ tests. My young cousin Rosalind, aged 13, did rather well: her IQ was 160. I also recall a bit of a problem with my experimental reports. The details are vague, but I remember that Peter Wason, a psychologist subsequently distinguished as the originator of the *Wason selection task*, was present at the painful interview. Peter Wason died in 2003. His obituary in *The Guardian* celebrates him as "Psychological pioneer of human reasoning who rescued his subject from the dark days of behaviourism." His rescue might have been even more successful had he urged my tutor to fail me. As it was, I scraped by the laboratory part of the course.

I managed to pass the internal exams in psychology after the first year which allowed me to go on to the second. But then there was a problem. I was called in late in the year to an interview with Dr. Kelvin, one of the professors. He told me that I was in trouble because I had failed my economics major. This meant that I would be out of the university – *finito!* "But I passed!" I wailed." You never went to the tutorials" he said. True, but I *had* passed the exam. Kelvin wanted to sack me, but he had chosen the wrong basis. He could have done so for any number of other reasons – not attending any lectures, or hardly any, not attending the economics tutorials, and so on. But, because of my switch from internal to external exam, he missed his target. After this mistake, he had to renege and I was allowed to continue into my final year. But the writing was on the wall…I was having fun seeing movies, writing for *Sennet*, checking out coffee bars, the new Italian-inspired venue for the non-alcoholic, hip student. With gleaming new Gaggia espresso machines (mysteriously labeled "Il funzioni senza vapore") and cool new music. I spent more time in them than in lectures or even, possibly in lectures + study. But could this go on?

In the summer of my first college year I rode the Vespa to Italy. I had a companion on the way down, Kenny, a Marxist-Leninist, who had stayed briefly in Cedar Road. He even looked like Lenin – had a pointy beard, wore the outfit and had the hat. He also hung out with Soviets, one of whom visited and played chess with us. I think the Russian was looking to recruit spies. But our indifference to politics soon put him off (we were founder-members of the Apathy Society but, naturally, had done nothing about it). The Russian may have been misled by a large ink drawing of the NKVD that I did and hung on the wall. And everyone was fascinated by the case of the self-confessed atomic spy, Emil Klaus Julius Fuchs, convicted a few years earlier.

Kenny was older than we and had spent some time in the Middle East. I am not a gourmet, but I still remember an extraordinary Arab meal that he prepared for us. Kenny's other gift (I assume it was his, I would never have chosen it) is a copy of Peter Kropotkin's wonderful *Memoirs of a Revolutionist,* which he left at Cedar Road. I carted the book, unopened, from home to home for fifty years but only read it recently, when I noticed that Kenny had 'liberated' it from the UCL library. The book is pre-soviet, pre-Lenin and pre-gulag, so it presents an idealistic vision not so much of communism as of a benevolent anarchism. Kropotkin was a Prince and his accounts of the Russian aristocracy and their relations to the serfs – who were treated more or less as slaves – is quite extraordinary.

In any event, for some reason Kenny wanted a lift to the South of France so we both set off on the Vespa for a journey absurdly long for that mode of transportation, as I eventually found out.

I dropped Kenny off in Cassis on the Mediterranean coast. The beautiful weather made a big impression. "Why, most of the people in the world live in climates like this" I thought, "who needs England!" Then I drove to Nice where I picked up Irene, my first real girlfriend. Irene was working in London, so could not take extra time to ride down with me. She instead flew to Nice where I picked her up at the airport. I had met her earlier, while still in SMGS, through the Young Conservatives, an organization I chose not for its politics, in which I was totally uninterested, but because of its social events. I particularly enjoyed the 'rambles' following well-mapped group walks through the countryside. It was on one such ramble that Irene and I first got together.

After I picked her up at the airport, we continued on the Vespa, along the French Riviera until we reached a little town on the Italian Riviera called Diano Marina. Looking around for a hotel we finally found a good deal in a new one still being built. The room was very nice, with a balcony. Most of the public space was unfinished, but there was a bar and there we were fêted by the locals. No idea why, but they were extremely friendly (possibly Irene – or the Vespa, an Italian icon – were factors). They could see we had no money and were very kind.

The Vespa finally quit on the way back. It could only be driven in first gear. I learned later that something called the gear-selector spider had broken. The 'toy' (as we called it) failed in the tiny little French town of Saint-Auban in the Alpes-Maritimes. I had to leave it for the AA insurers who – eventually – brought it back to London. But we had almost no money to pay for a return train fare. Somehow we made it to Lyon where I got a loan from the British Consulate. It was the only option in an era before credit cards, and my parents in Africa totally out of reach.

Life in Cedar Road resumed. The house was fully occupied: three of my friends, plus the secretarial lady and two American year-abroad visiting students and of course my grandmother. I was not impressed by the Americans. They seemed to have been mightily spoiled by their parents. Messier than the Brits and (it seemed to us) less mature. But maybe their colleges were to blame: one kid was from Oberlin, the other

Antioch. Of course, they attended lectures more conscientiously than we did because in the American system attendance was mandatory.

Left: Tony Stead. *Right*: Artist at work: Brian Taylor. Both ca. 1954. *Middle*: JS cruisin' in the Midlands' and Brian, Bob W. and Tony at play.

My three friends were Tony Stead, Brian Taylor and Bob White. Bob eventually dropped out of our little circle, but Brian and Tony I know still. Tony is from Bradford, where his father ran a DIY shop. He was a good-looking lad with great aptitude at sport. Not that he ever played any. But I remember the first time I introduced him to tennis – at which I was mediocre but which I did play occasionally. He was amazingly good and half-volleyed like a pro. Tony was also exceedingly bright, especially in math. But also very idle, spending much of his time playing, and winning, at poker in the UC lounge. In the end, for lack of a few weeks study he failed his maths course – but went on to success with the advertising firm J. Walter Thompson in Berkeley Square. There he ran a research operation and invented the Yorkie bar. Who among us can boast a nobler achievement!

I first encountered my artist friend Brian (not the knife-thrower Brian!) when I was working for the Post Office at one of London's big rail termini. Students were added to the workforce to cover the Christmas rush every year. The enthusiasm of the toilers was quite variable. Mailbags would be stacked in huge volcano-like piles and often, anticipating *I'm All*

Right, Jack[5], in the crater were to be found workers concealed from view and playing cards. Such was the British work ethic in 1955. I first saw Brian, in his trademark paint-stained (but expensive) gabardine raincoat, with his arms in the air gesturing at all the students dragging mailbags like insects with their pupae: "Ants! Fucking ants!" he exclaimed. How could you not love a guy like that!

The Post Office was only one of several part-time jobs I had in those first two college years. I worked one summer on the watch counter of Selfridges in Oxford Street pretending an expertise that I mostly lacked. Better jobs involved driving. My uncle Eric had taught me to drive his new Vauxhall and I got my license from the London Country Council. Driving was a skill not widely shared in those days, so I was able to get part-time jobs. One was driving a limousine for weddings. It was a gleaming, but ill-maintained model with the wonderful name of Humber Super Snipe. I remember a rather raucous Irish wedding with the car full, not to say overfull, of well-lubricated guests. All went well until we got to a hill, up which the Super Snipe would not go. The clutch was slipping. The Irish were unfazed. They got out and started to push the car and all went off without further incident.

I am sorry I did not get to know my uncle Eric a bit better. I saw him rarely; and our conflicting diurnal rhythms (he was always on night work) didn't help. In his youth, when he had more time, he wrote two novels: *The Hamadryad[6]*(1936) and *The Pagoda Tree* (1939) neither a great success, but a sign of promise. He was very interested in Indian history. Given time, he would have written about it. I do remember once when I was about twelve, being taken by Eric to a very fancy lunch at The Connaught, an Edwardian-style eating establishment that still exists. Jacket and tie were required. (But that attire was normal in those days. Photos show me thus clad even on holiday.) I retain an image of much polished wood and brass, and a round of beef carved in front of you.

On another occasion I was driving a delivery van. I made a call at the Rolls-Royce Park Ward coachwork factory which was then in Willesden. I was delivering foam-rubber footrests (!) for a R-R Continental they were working on that was to go to Mr. Jack Warner, the American movie mogul. Apparently Warner was at that time engaged in protracted negotiations with Winston Churchill over the movie rights to the first volume of his autobiography: *My Early Life*. The negotiations eventually failed and it was only after several years of back and forth with potential

---

[5] The classic 1959 Peter Sellers and Ian Carmichael movie that satirized both workers and management in industrial Britain.

[6] A mythical Greek tree nymph, but also the name for the King Cobra snake of India.

financers, producers and writers that a movie – *The Young Winston* (1972) – finally emerged[7].

Friend Brian Taylor came from an upper-middle class family somewhere on the Welsh border. His father was something on one of the Manchester exchanges. His mother was a compulsively neat housekeeper. Brian was not. He was a dedicated artist, studying at the Slade School, which is part of UCL. Brian acquired a girl friend who seemed much younger than us. We accused him of baby-snatching. But of course she wasn't that much younger, since we were all just twenty or so. Stephanie and Brian are still together and Brian is still painting.

As far as I know he has never had a regular job, but he did have one great commercial success. In collaboration with one of the entrepreneurial

Brian Taylor, self-portrait in 1986.

Forbes family, in the late 1970s, he invented a board game called *Kensington*. It was packaged like a long-playing record, and was a great hit. I used to watch the Johnny Carson late-night show. As the show began one night, I got a surprise phone call. "I'm on the fucking Carson show!" came the voice. It was Brian promoting *Kensington*. A few minutes later he appeared on the Tonight stage in a scruffy oversized sweater, which he offered to remove in response to a crack from Carson.

Whether the engine was he or Peter Forbes, the publicity for *Kensington* was brilliant.

Brian stayed with us in North Carolina while he was on a promotional tour of U.S. radio stations and we started to write a tell-all book about the tricks he and Forbes played to get publicity. But I think they decided it told a bit too much and nothing came of the project.

Brian, Tony and I had a holiday at the end of my first college year. Brian's father owned an old 30ft or so motor launch, docked on a canal in

---

[7] The ebbs and flows of Churchill's finances and his intricate negotiations with publishers, movie producers and the Inland Revenue, all interwoven with a taut summary of his life, are unraveled in David Lough's brilliant *No More Champagne: Churchill and His Money*. Picador 2015.

the midlands. Brian loves canals so we went on a little excursion in the boat. I found that steering something that turns from the back rather than the front is not easy. I learned about locks, the inevitable accompaniment of any canal trip. We wended our way from the outskirts of Manchester to Warrington. The high-point of an unglamorous adventure was a truly awful movie comedy, *Fun at St. Fanny's,* we watched on the way.

Brian's devotion to his painting is unwavering. Most amazing of all he never turned his considerable marketing and publicity talents to promoting his own art. It is a fact generally acknowledged that most successful artists are second, if not first, brilliant self-publicists (Hearst, Emin, Koons, q.v.). Brian chose just to paint and make a few quid some other way. Our house is full of his restful Cezanne-like landscapes and *natures-mortes* – and a splendid self-portrait.

I acquired a new girlfriend sometime in my second year at UCL. Patricia was a fellow student, much more conscientious than I and in a tougher subject, mathematics. She was always elegantly dressed – nothing like a student! – with a rather aloof (shy?) demeanor that concealed great kindness. And she had a car, unheard of for a student in those days. She was in fact from a rich Jewish family in the wool business with factories up north (Bradford, I think). The family owned a substantial property in New Oxford Street. We got along very well and she invited me to a party she and her brother had organized in what seemed to me an amazingly opulent family apartment in a block of flats north of Hyde Park. (The building – or at least, my memory of it – reminds me of Hercule Poirot's residence in the BBC series). It was around Christmas time. My memory of her parents' flat is: *gold.* When she went on holiday with the family I would get sweet letters from her with letterheads like Hotel Bristol, Nice, Hotel Normandy, Grande Plage Deauville and Lido Venezia. We saw each other briefly when I returned from Rhodesia, but then lost touch when I went to America.

Contact was resumed a few years later. I had long been a member of the Royal Institution in London, in memory of their Christmas Lectures which I always much enjoyed as a kid. Still in a grand Palladian-style building in Albemarle Street, just off Piccadilly, it is an organization with a wonderful history, home in the 19th century to Michael Faraday, electrical pioneer, and Humphry Davy, discover of sodium and potassium – both great scientists and magnificent lecturers. In the twentieth century it housed Nobel-winning crystallographers W. H. Bragg and his son and, for a while Desmond Bernal, of whom I have just written. Most recently, somewhat damaged, both financially and reputationally, by the baleful influence of its recent director, the bizarrely successful lightweight Baroness Susan Greenfield, it became a sort of scientific Kardashian. The RI is slowly recovering after sacking the Baroness in 2010. The last time I saw the Prof. Greenfield, a neuroscientist by training, was in 2003 when she gave an invited talk at York University, where I was spending a sabbatical leave. The talk was superficial, and not even slickly presented. She showed slides – yes actual slides, not I think, PowerPoints – dusty with age, to make unoriginal points. How and why this lady achieved such influence in the U.K. is a mystery.

In 1973, while I was on sabbatical leave in Oxford, I saw that Dr. Edwin Land was to speak at the Royal Institution. He is famous as the inventor of Polaroid photography and a great contributor to the theory of color vision. The RI is known for its scientific theater, and I expected Land to wow the audience with demonstrations of color perception from black and white slides illuminated with only two colors. He had shown that people will see a full range of colors in a superimposed projection of two black and white slides of the same scene one, one taken with no filter and the other with a red filter. All should appear pink, but in fact people see other colors like green and yellow, albeit rather desaturated. These experiments of Land's evidently grew out of his attempts to understand early work on color vision by the great 19th century physicist James Clerk Maxwell, the man who provided the theory that explained Faraday's brilliant electrical experiments. I decided to take young son Nick to see Land. After the talk, which delivered the goods, in the crowded lobby I saw Patricia, with her husband Jimmy, whom I had never met. They were both lovers of science. Patricia got Jimmy a birthday present of 16 original etchings from Johannes Kepler's 1609 *Astronomia Nova* a few years later, I remember. They were also strong supporters of the RI and attended the talks regularly. So our friendship resumed and we usually see them both when we visit the U.K.

Finally, a crisis I might have anticipated given my grandmother's loss of an eye a few years earlier. One afternoon late in the winter of 1957 I planned on riding the Vespa the 60 or so miles to Cambridge to visit Ivan and stay the night – a rare night away from home. My grandmother was not feeling well, but urged me to go. When I got to Cambridge – it was dark, so must have been well into the evening – I decided I must phone home to see if Gran was OK. A long-distance call was not an easy thing in those days, but I found a phone, dialed – and got no answer. There was nothing for it but to ride back 60 miles in the rain. I found an empty house and a note saying that Gran was in the hospital. She had had a stroke but somehow managed to contact her son Eric in Hampstead, who had her taken to the hospital. She never returned to Cedar Road. After she somewhat recovered from the stroke, she stayed with Eric for a while and then went to a care home in Golders Green where she lived on for many years.

So, I survived – just – my first two college years. But all was not going well. I had not found a research topic that excited me; my journalistic excursions were fun but had not set the town afire; and the drabness of Britain in those days was beginning to get me down. My parents were back in the U.K. on my father's first leave. Going back to Rhodesia with them and driving my Dad's new tax-free car from Cape Town to Fort Rosebery was an enticing prospect. I decided to return with them to Rhodesia, planning to return to UCL after a two-year leave of absence – with more money and, perhaps, more sense.

News item you might have missed...

HARVARD PRESIDENT'S REMARKS CONDEMNED

At a recent conference on women's representation in sports such as weightlifting, football and caber-tossing, Harvard President Lawrence Summers was asked to comment on the economic and scientific aspects of the problem. Asked to be "provocative" President Summers (speaking from notes) said that the under-representation of women in these sports may not be entirely due to discrimination. He pointed out that tests have shown women to be, on average, smaller and physically weaker than men and so less able than men at some sports. He added that "Some researchers have suggested that these differences may be biologically based."

Gasps and groans from the largely female audience greeted these remarks. Triathlete Imelda Oxblood commented "I was shocked! I found it extremely upsetting that someone in Summers' position could still advocate these tired old stereotypes." She left the hall in tears, adding "I simply couldn't stand to hear any more of Summers' sexist ranting!" Another woman in the audience commented "Just remember: Once they said that women couldn't drive cars or even read. Surely we've moved beyond that." Moderator Tawana Mbeki reminded the audience that at an earlier conference Summers had made similar comments. "He said that African Americans are darker skinned than whites!" she said. "And he was on this biological kick then too!"

Today at a press conference President Summers apologized for his remarks, which he called "insensitive": "I'm afraid I was not fully aware of all the social-science research on this problem" he said. He agreed to convene a task force of experts, members to be democratically chosen from the Women's Sports Collective and the Union of Concerned Women at Harvard, to provide an objective report on the underrepresentation issue.

Summers' office declined to comment on a report that President Summers had agreed to appear at the annual Hasty Pudding fancy-dress party dressed as Eleanor Roosevelt.

Chapter 4: Fort Rosebery

My parents and I set off in September 1957 on the Union-Castle vessel *Edinburgh Castle*. My sister Judy had flown back earlier to her school in Lusaka. We left from somewhere in the 26 miles of London docks that then existed. Now all are gone, replaced by expensive residential real estate. If you want to catch a ship from London now you have to go to Southampton or Harwich. The voyage was pleasant but uneventful. I won something at bingo – the only time I have ever managed to win a gamble. I saw the King Neptune ceremony to induct passengers crossing the equator for the first time. When we stopped, I admired the colorful town of Las Palmas in the balmy Canary Islands. I glimpsed Tenerife – famous from Darwin's *Beagle* voyage and for Wolfgang Kohler's experiments with apes.

We got to Cape Town after some two weeks at sea. It was and is a beautiful place, a splendid bay overlooked by a ring of small mountains. Whitewashed houses with red roofs: All was bathed in sunlight. I still remember the taxi we took from the dock – an American Rambler. The size (small for the U.S., large for us) and incredibly soft suspension made an impression. We stayed in the King Arthur's Seat hotel for a night – many black and 'colored'[1] servants and modest luxury. We set off the next day in Dad's new white Ford Consul (4-cylinder, we couldn't afford the 6-cylinder Zephyr). The car had no air-conditioning, but it did have a visor that shaded the front window and a sort of Venetian blind for the rear.

The first event of note was a speeding ticket near Paarl, home of what the English South Africans called derisively the 'Paarl bray', the strong Afrikaner accent of the locals. The cop had the accent. The ticket was my fault. Excited by a new car and an empty road, I drove much too fast – which led to mechanical problems later, I suspect. It struck me even then that speeding is what would now be called a 'pre-crime', after the Tom Cruise movie *Minority Report*. People generally speed when there is little traffic, the weather is good and there is a clear road, i.e. when it's safe. No harm is done. But that's also when the cops get them. After all, the speeder *might* cause harm later. Pre-crime, then.

We crossed the Karoo, a dramatic semi-desert just north of Paarl, then on to Louis Trichardt and across Beit Bridge into Southern Rhodesia (now Zimbabwe). The Limpopo River was not "grey-green, greasy, all set about with fever trees" but rather muddy. There was fever about, and not just in the trees, as I was find out.

---

[1] Mixed race.

SR was then the second most prosperous country in sub-Saharan Africa, right after *apartheid* South Africa. It had a vigorous farming economy, tobacco and coffee – export crops – but also maize – 'mealies'). The farms were mostly run and owned by white farmers. Now, after some 35 years under racist thug Robert Mugabe, abetted by a Labour government that in 1997 reneged on the 1979 Lancaster House agreement which might have protected them, the white farms – and their black farm laborers – are almost gone. Zimbabwe is second from the bottom on the African economic ladder.

COLONIAL AFRICA

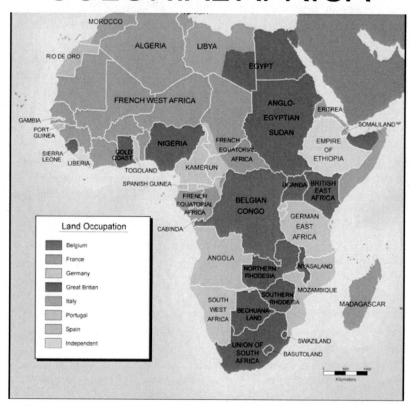

We drove on to Salisbury (now Harare) capital of Southern Rhodesia. The roads now were no longer metaled. They were 'strips'. Instead of tarmac across the whole road, two narrow strips, a car-width apart, served as an improvement over dirt, which was smooth when graded but acquired nasty traffic-induced corrugations after a while, even in the dry season. Of course, there were only two strips, so every encounter with an oncoming vehicle was a game of 'chicken' – who gets off the strips first.

Onwards into Northern Rhodesia and its capital Lusaka. Northern and Southern Rhodesia and Nyasaland (Malawi) were then all linked into the Central African Federation, so travel between them was easy. A very few years later the CAF was fiercely criticized from all directions and was soon to be abandoned by Britain. Even South Africa, still under apartheid, was keeping its distance. English doctor Theodore Dalrymple arrived in SR just before the CAF was destroyed:

> I expected to find on my arrival, therefore, a country in crisis and decay. Instead, I found a country that was, to all appearances, thriving: its roads were well maintained, its transport system functioning, its towns and cities clean and manifesting a municipal pride long gone from England. There were no electricity cuts or shortages of basic food commodities. The large hospital in which I was to work, while stark and somewhat lacking in comforts, was extremely clean and ran with exemplary efficiency. The staff, mostly black except for its most senior members, had a vibrant esprit de corps, and the hospital, as I discovered, had a reputation for miles around for the best of medical care. The rural poor would make immense and touching efforts to reach it: they arrived covered in the dust of their long journeys. The African nationalist leader and foe of the government, Joshua Nkomo, was a patient there and trusted the care implicitly: for medical ethics transcended all political antagonisms[2].

The CAF collapsed in a storm of anti-colonial propaganda and racial acrimony in 1963 – a tragic and unnecessary ideologically motivated destruction of a federation that might have evolved peacefully into something better. Instead, it was allowed to degenerate dangerously into something much worse.

We visited my sister Judy, who was at the Jean Rennie government boarding school (renamed Kabulonga since independence) in Lusaka. The next step would normally have been to go on to Ndola and the copperbelt and finally into the Belgian Congo, crossing the pedicle and the Luapula River via the Chembe ferry and on to Fort Rosebery. But this time, the pedicle route was not possible, because the Luapula was in flood and the Chembe ferry not operative. So we had to go the long way round, through the Northern Province and Kasama.

I made the return trip, from Fort Rosebery to Ndola and back, once later with a friend. That time, we went the long way, via another ferry, at

---

[2] Oh to Be in England *City Journal*, Spring 2003 Vol. 13, No. 2. Theodore Dalrymple is the pen name for physician Anthony Daniels.

Kasenga. On the way back we got through the customs house at the Congo border, only to find that the Kasenga pontoon ferry, across the Luapula, had closed for the night. So we were trapped between the border office, now closed, and the ferry, also closed. We had to sleep in the car. It was too hot to close the windows. I counted some 30 mosquito bites on the back of one hand the next day. Not a good night.

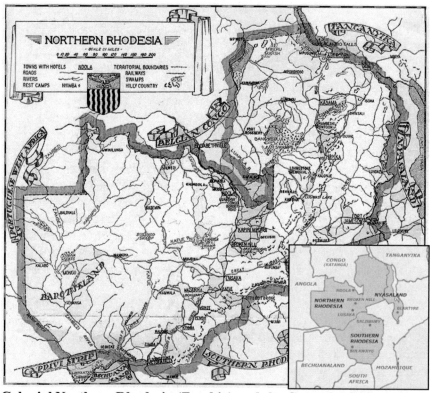

Colonial Northern Rhodesia (Zambia) and the Central African Federation: NR, S. Rhodesia (Zimbabwe) and Nyasaland (Malawi) in the 1950s.

This time we made it to Fort Rosebery without incident. We came to a nice government house with a biggish garden, containing a large termite mound (on which I saw a leopard, one night much later). No electricity – lighting was via kerosene[3] lamps, either Tilley or Aladdin. And of course no telephone. There was a small fridge, a no-moving-parts, kerosene-powered Electrolux. The windows all had fly-screens – routine in Africa and America, but a novelty to a visitor from England. But we did have a

---

[3] Kerosene (U.S.) = paraffin (U.K.).

couple of good servants who lit the lamps and did the cooking. The roof was aluminium and the ceilings some kind of white fabric. There were only 22 'European' families in Fort Rosebery.

Sitting on the front 'stoep' (Afrikaner word, of course) one evening I was surprised to see hundreds of bats emerging from the corner of the roof in a great 'whoosh'. They lived during the day in the space above the ceiling and left to forage at a precise time every evening. Once a young bat slipped through a gap between ceiling and wall and flew about the living room. I eventually caught it, whereupon (naturally) it bit me. Whether I was unaware of, or didn't care about, the risk of rabies, I don't remember. But I did nothing except put a band aid on the cut and all was soon well.

Top: **Our house in Fort Rosebery (termite mound to the right). Note the shiny aluminum roof.** *Bottom left:* **Downtown Fort Rosebery, ca. 1958.** *Bottom right:* **On the way up from Cape Town.**

Northern Rhodesia then was a protectorate, not a colony. British settlers were no longer allowed and the law made it clear that N. Rhodesia was eventually destined for home rule. There were a few early settlers, though. Sir Stewart Gore Browne in 1920 built an estate with a grand house to the east of Kasama in the Northern Province called (after a nearby lake) Shiwa Ngandu. The house is very much in the English manorial style, with walled gardens and tennis courts. "The estate followed in the tradition of 19th century utopian model villages like

Saltaire and Port Sunlight...[it] had its own schools, hospitals, playing fields, shops, and post office. Workers lived in brick-built cottages and the estate was ruled as a benevolent autocracy..."[4] Gore Browne tried various crops, with mixed success, but through it all came to understand the local Bemba people pretty well. When he died in 1967 he was given a state funeral by Kenneth Kaunda's independent Zambian government (quite a contrast with white-hater Mugabe, although Kaunda also expropriated a few white properties). There was also a connection to Fort Rosebery, which I'll get to in a moment.

My sister's Lusaka school friend Megan later married into the Landless family who for five generations have farmed in an area called Landless Corner, north of Lusaka. They still run their farm feudal fashion, much as Sir Stewart ran his, providing lodging, food and medical care for several hundred farm workers and their families. The last time I saw Megan, a few years ago, in England, she had just finished supervising the delivery of worm medicine for the entire crew.

Left: **Sister Judy, aged 16 in the garden of our house in Fort Rosebery, in 1958.** *Right:* **Mum and adored dog Blondie.**

The next thing was for me to get a job. I applied to many employers, ranging from the Provincial Government, the Ndola Copper Refineries Limited, Rhodesian and General Asbestos Corporation (Private) Ltd., Rhodesia Mercantile Holdings (Pvt.) Ltd. (whatever that was), the Rootes Group (cars), the CAF Federal Broadcasting (Junior Announcer/Producer) and several others, all with no success. I eventually found work right in Fort Rosebery with something called the Health and Nutrition Scheme, which had just been started. It was headed by an Irish researcher who was

---

[4] http://en.wikipedia.org/wiki/Shiwa_Ngandu

soon joined by a Danish medical doctor. Both were supported by the World Health Organization. The rest of the operation was paid for by the Northern Rhodesia Government (NRG).

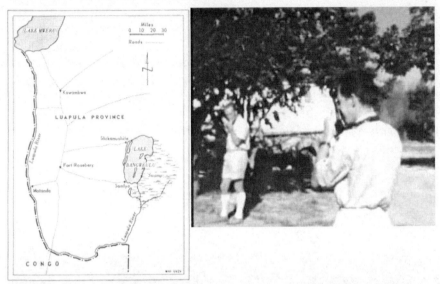

Left: **The three study sites of the Health and Nutrition scheme in Luapula Province in the 1950s: river site at Matanda, lake site at Shikamushile. The main site and lab was at Fort Rosebery.**
Right: **Fergus McCullough on the left, Bent Friis Hansen on the right, at FR about to set off to a study site.**

The Scheme was intended to study disease – malaria and bilharzia (schistosomiasis) especially – and nutrition, especially *kwashiorkor* a protein deficiency that affects development and leaves kids with skinny limbs and bloated bellies. The scheme had three study sites[5]. The main site was Fort Rosebery, which had a laboratory and a small hospital. The other two sites were Matanda on the Luapula River and Shikamushile on Lake Bangweulu. Each was a one-hour or so drive from FR on dirt roads either in a Land Rover or a bone-rattling pale-blue Bedford pickup ("vanette") that seemed to have no suspension at all. The contrast with an American-made Dodge I drove later was striking. Either Americans all have weak spines, or the Brits don't know how to design a suspension.

The senior researcher on the scheme was Fergus McCullough, an Irish expert on tropical diseases who worked most of his life in Africa. Fergus

---

[5] See Fergus McCullough & Bent Friis-Hansen (1961) A Parasitological Survey in Three Selected Communities in Luapula Province, Northern Rhodesia. *Bull. Org. mond. Santé,* 24, 213-219, which is their report of the Luapula research.

was a rather charismatic figure. Charming, tall, bearded and athletic, and a bachelor, he made waves among the young women, married as well as unmarried. Alas, I never found out the details. Fergus was a great tennis player and a kind man – kind enough to coach me a few times. And kind enough (unlike the Danish guy) to feign interest when I expounded on the correct way to set up a microscope.

Soon after I returned to the U.K., Fergus finished up in Fort Rosebery and went on to another project in Ghana. He there married a feisty Irish woman he met in Belfast while finishing his thesis. He died, apparently of the same malaria-induced condition as my grandfather, in 1995, at the early age of 69. All this is well told in his wife Elizabeth McCullough's book *Eighty Not Out: Memoirs of a Bad Mixer*[6].

Health and Nutrition and the staff. Fergus McCullough is on the right. JS taking picture. Bedford "feel the road" vanette (pick-up truck) behind – on our way to one of the study sites.

The Danish guy was a pediatrician, Bent Friis Hansen, who fit most of the Scandinavian stereotypes. His job was to examine hundreds of children for disease and malnutrition. He left for Copenhagen before the end of 1959 but we corresponded for some time about projects that I worked on in his absence.

I was hired as a lab technician and physician's assistant. My only qualifications were (1) I was there; (2) I had some scientific training; (3) there was (probably) no one else. I worked both in the lab and as a note-

---

[6] Belfast: Blackstaff Press, 2012. 'Mixer' is a reference to the alcoholism she eventually overcame.

taker for Friis Hansen. I spun in a centrifuge test-tubes of blood collected from hundreds of children in the three study areas to get a hematocrit (packed cell volume) measure of anemia. We got a Beckman instrument – rather hi-tech for its time, and quite expensive (£900 according to my notes). It could make spectral measurements on blood and urine samples in square quartz cuvettes. We collected urine specimens from the kids. I can still see these little bottles arranged on a shelf: yellow, pink, red, red... – the red and pink, from blood in the urine, indicating bilharzia infection, which was endemic, especially in the river site. But not so much in the site on Lake Bangweulu.

There are two kinds of disease: diseases that kill quickly, if they kill at all, and diseases that debilitate and if they do kill only do so after a long time. The latter are of course much worse for the economic health of a country. Ebola and one kind of malaria are diseases of the first kind; bilharzia is a disease of the second kind. It weakens its sufferers but rarely kills them. Bilharzia is a singularly charmless affliction. It is caused by a parasitic, trematode worm about 1 cm long that lives in your veins and subsists on red blood cells and other things essential to life. The worms are monogamous. They come in pairs and reproduce as busily as they can in your body. Since the worms don't multiply in the body, the amount of sickness depends on the patient's fixed 'worm load'. The eggs, when they don't get lodged in an organ and cause trouble that way, are excreted in the urine. If they're lucky they end up in water – a river or lake –where they hatch into tiny larvae, which go looking for snails. When they find the right kind of snail, they bore into it and after a few weeks produce sporocysts which each produce thousands of *cercaria*. These little darlings then swim about looking for bathers into whose skin they can burrow. There they turn into the trematode worms and repeat their sorry cycle.

Malaria and bilharzia were the main diseases affecting the local Bemba people. Friis Hansen examined hundreds of children while I took notes and the African staff collected blood and fecal and urine samples. I still remember some of his more frequent diagnoses: "Sway back one plus", in Bent's Scandinavian accent. Sway back is a small deformation of the spine, one-plus being the severity according to Bent's scale. *Gynecomastia* – small breasts in boys – harmless but indicative of other problems, was also quite common. Weight and skin-fold measurement revealed malnutrition.

Friis Hansen almost killed me on one occasion, although I daresay that was not his intention. We were all camped at one of the study sites and Fergus McCullough was supposed to come up from the Copperbelt to join us for supper – a drive of some 200 miles over crappy roads. The sky was clear; it was the dry season. Luapula Province is 4500 ft. above sea level so the nights were cold, even though we were in the tropics. The only heating was a campfire. Bent insisted that we all wait to have supper until Fergus arrived. Unsurprisingly, he was late, not arriving until near midnight. Bent, just come from Denmark, was unfazed by the temperature which, to him, must have seemed pleasantly mild. Not so for me. I weighed at that time something in the region of 126 lb, a bit on the low side for someone over six feet tall – no insulation. I had been in Rhodesia for several months. I had no warm clothes. I was freezing.

When we got back to Fort Rosebery the next day I was seriously sick. An X-ray (at a dose well above what would now be regarded as a safe level, I am sure) showed a spot on my lung. The diagnosis by our not-very-good local Brit doctor, was viral pneumonia. I was sick with very high fever for about a week, my temperature was brought down only when one of the nursing sisters (Rosanna? God bless her!) on her own initiative gave me an anti-malarial, even though I hadn't been diagnosed with malaria and was taking Paludrine, the only prophylactic then available. When I came home I was feeble and could walk only a few feet at a time. I was OK just sitting down, which I did a lot. It took a few weeks to get back to normal.

The only thing I remember from the time I was sick was sitting on our little stoep when two young white guys in white shirts and ties came up from the road to chat. They were of course Mormons (or maybe Jehovah's Witnesses?). In any event, I was still weak and moving was an effort so we had a nice talk about spiritual matters. I could not convert them, and vice versa. I was impressed that they would venture so far to promote their faith, improbable though much of it seemed to me. Maybe they had better luck with the Bemba.

There were several religious groups in the area. The White Fathers were a French Catholic Missionary order, so called because their dress was a white burnoose. The nurse-sister who saved my life belonged to a related order, I believe. Another group was at the Mansa Mission – *Christian Missions in Many Lands* – an offshoot of the Plymouth Brethren. The Mansa Mission people were notable for their involvement with the Bemba. Two missionaries G. W. Sims and Wm. Lammond, studied Cibemba which, until their work, was not a written language. George Sims produced the first Cibemba grammar, *An Elementary Grammar of Cibemba*[7], which was dedicated to Lammond:

DEDICATION

---

[7] Published by Mansa Mission Christian Missions in Many Lands, Fort Rosebery. (1959, Morija Printing Works, Basutoland).

Chapter 4: Fort Rosebery

To Wm. Lammond Esq. M. B. E., to whom the writer is indebted for his first lessons in, and love of, the language of the Bemba people.

I recited this example of white affection for tribal Africa during a debate (on colonialism or some such topic) in front of an unsympathetic audience at Duke University many years later. My opponent, a cultural anthropologist whose current interest is masculinity and Native Americans in advertising but who had some experience in Africa, grabbed my copy of the book and after much leafing through it found among the "Exercises", an instruction to servants. "Aha!" he exclaimed "Listen to *that!*" to general applause. The racism of the imperialists was confirmed, not that this audience needed any new evidence.

I looked again recently in Sims' book for the cited passage, or one like it. I found many innocuous phrases like this: *bushe abana bacili kwi sukulu?* (Are the children still at school?): I could find nothing that corresponds to my memory, nothing corresponding to even remotely to my colleague's claim. Perhaps he would have considered this racist: *Babwana bali kwi nomba? nalimo bacili ku Kitwe.* (The Bwanas are where now? perhaps they are still at Kitwe?) or perhaps this *twala bafibashi ku cipatala kuli doctor, takuli muti kuno* (take the lepers to the hospital to the doctor, there's no medicine here).

Much of the administration of Luapula Province was carried out through District Assistants, young men, often with no college training, but from 'good' schools, or with military experience. The government required them to learn the local language and gave a pay boost if they passed the relevant examination. Even non-administrative government workers – like me – could compete. The reward for passing the exam was £20, a considerable sum in those days. So I took a course taught by one of the Mansa Mission people and passed. Alas, I now remember only a couple of words of Cibemba.

By happenstance, decades later my sister re-connected with one of the DAs, Alan Kitching, who lives not far from York, our summer retreat in the U.K. And very recently, at a Durham North Carolina Rotary meeting, I was introduced to a young Bemba-speaking woman, a Rotary Peace Scholar arrived from Zambia just hours before, who was delighted to be greeted "*Mwapulene mukwai*" – "how do you do" – one the of the very few Cibemba phrases I could remember.

Four blurry screenshots from an 8-mm movie. *Top left:* **Fergus McCullough talking with Betty Thompson at the beginning of a trip to Matanda on the Luapula River.** *Top right:* **Mothers and children being delivered by vanette to the FR lab for testing.** *Bottom left:* **Lady Lorna Gore-Brown canoeing on the Luapula.** *Bottom right:* **Lady L's Luapula hut.**

Racism – its evils, its ubiquity, its ineradicability – has become a neurosis of the Western world, the U.S. especially. The *New York Times* can write of little else. So, how racist was colonial Northern Rhodesia? Not very, considering. The Mansa Mission folk set the tone by their great interest in the local people. There were no racial laws, like S. African apartheid. There was no legal segregation. Criminals were tried in both British and tribal courts so there was some local autonomy. Accused generally preferred to be under the British courts, or so I heard. On the other hand, Fort Rosebery had a little club with, as I recall, only European (i.e. white) members. But the club was very modest in its facilities: tennis courts a small library and a bar, basically. I don't think there was any demand from the locals to join – nothing like the toxic situation that George Orwell imagined in *Burmese Days*. Among many without first-hand experience of Africa, Orwell's vision has prevailed, I fear.

So were the Africans treated as equal to the Europeans? Well no, of course not. The Bemba culture was pre-literate, most local Africans were completely uneducated, in a Western sense. Tribal Africans scored abysmally on standard IQ tests, even nonverbal ones. The Bemba might well have been much cleverer about local matters, but the whites had no way to see that.

Superstitions were rampant among the locals. They were very afraid even of some harmless wildlife, like lizards, for example. I found this puzzling: why weren't a people presumably accustomed over many generations to living in the bush apparently rather ill-adapted to it? Health and Nutrition gave rise to its own superstition: a rumor that the pathogen-filled blood samples we took from children at the study sites were being sent to England to be somehow exploited (drunk?) by white people. And of course most of the blacks that the Europeans encountered were servants, usually men – or boys, like the 'ball boys' who retrieved on the tennis courts. (Pay for the ball boys was one 'tikki' – three pence. To be more generous was to risk social sanction.)

I worked with many native medical people at Health and Nutrition lab. They were much better educated than the average. But to a half-college-educated, young and totally-inexperienced-in-the-ways-of-the-world Brit they seemed rather simple. Whether this was accurate or simply represented a protective pose I don't know. The result, inevitably, was that whites regarded the blacks as almost like children. But there was little animosity – and some affection – on either side. My father was especially popular with his workers. I still have an 8 x 10 portrait photo of one of his servants, Wilson Nyonyo Thole, that he was given as a friendly token when he left. This amiability was to disappear a year or two after my own departure. Agitation for independence led to anger and sporadic attacks, even in Luapula Province.

My sister Judy told me about one incident which, had it happened in segregated Alabama or Mississippi, might have led to violence. In Rhodesia, it was occasion only for amusement. Judy, then a leggy age thirteen, clad in tennis shorts, was with a friend in downtown Fort Rosebery in 1955 or so. A bunch of young African boys was hanging around nearby sniggering and pointing at the girls. "What are they saying" said Judy, to her White Rhodesian friend Jean, who understood some Cibemba. "That *muntu* [man] says he would like to marry you," said Jean, euphemistically. Thirteen was of course a marriageable age for Africans in those days. For the Bemba, bare legs were an invitation (bare breasts, not so much). So Judy was hot stuff. But no offense was taken.

After about a year Bent and Fergus decided to drive Fergus's new Ford Zephyr north for a holiday. They planned to visit some famous beauty spots: Goma, Bukavu and the Ruwenzori mountains in Rwanda and Uganda. They failed to invite me to come along, but they picked the right time. Thirty five years later this region was home to Idi Amin and a horrible genocide.

There was not much to do for recreation in Fort Rosebery. There was a little beach at Samfya on Lake Bangweulu about an hour's drive away. Those so inclined swam and sailed on the lake. The lake abutted a huge swamp and chunks would sometimes break off – this is how the swamp grew. One such 'floating island' obliterated Samfya beach for a while and had to be removed. There were occasional films at the club, but I don't remember any of them. The two main diversions were tennis and drinking. There was also a rather pathetic makeshift golf course, which I played on once. It put me off the game for life.

Boating on an African lake is not without its risks. A few years later a colleague of mine at Duke, biologist Dan Livingstone (appropriate name!), an expert on Africa's rift-valley lakes, was attacked by a crocodile while afloat on one of them. The creature bit his boat in half. He and his companion survived simply because the croc was more interested in the boat than in them.

A bad night on the pedicle. *Left:* **Alan Hill and his Humber Hawk on the Pedicle Road.** *Bottom:* **The Kasenga pontoon on the Luapula River, ca. 1957.**

I didn't drink, but I did play quite a lot of tennis – which led to my only sporting triumph. I played in a local tournament. In the mixed doubles I was paired with Betty Thompson, wife of the Provincial Commissioner Euan Thompson. Betty was a steady if unspectacular player. I was erratic. Our opponents were a young chap who had been a Junior Wimbledon contender and a lady whose name I don't recall. So we were the underdogs. But after a long battle, Betty and I won as darkness descended. I rested on these laurels for the rest of my sporting life.

Gardening was a favorite pastime for many in FR even though people knew they would live in their houses for only a few years before being transferred. The availability of 'garden boys' helped but the real key was 'black earth'. The red lateritic soil that covers much of the country is not fertile. But every so often as you drive along a rural road in NR you come across swampy areas known locally as 'dambos'. Dambos contain fertile black soil which, if deposited in your garden, will do wonders. Some government workers were rumored to use government trucks to smuggle in black earth for their personal use (shock! horror!). My father referred to one of them as "Mr. Tresham, MBE", where MBE stood not for 'Member of the British Empire" a popular 'gong' (decoration), but 'More Black Earth'. My father was somehow able to obtain an adequate supply of the magic substance and had a beautiful garden.

So, for recreation I read and took photographs. I got a nice camera, a Zeiss Contaflex, one of the very first single-lens reflexes, by mail from an Indian shop in the duty-free port of Aden in Yemen (£24! So much for my Bemba-language money). Yemen was then a relatively tranquil British protectorate. Now, of course, it is a violent mess. H & N had a darkroom and enlarger so I could develop, enlarge and print my photos. Color slides were also just coming into use.

This hobby led to some gainful employment. I took all the pictures for a little display that H & N put on to celebrate a visit from the Governor General, the gorgeously caparisoned Earl of Dalhousie. I was a bit grumpy that I was at first forbidden to get payment for them, even though the work was done after hours (government rules, apparently). After some negotiation, a small amount was paid, and I did get a nice letter from the Governor General.

The local people lived mostly on cassava (manioc). Not only is cassava poisonous until crushed and rinsed with water, it has almost no protein. So protein malnutrition was common. The Northern Rhodesian Government had many schemes designed to add protein to the Bemba diet: breeding pigeons and fish for example. These schemes were administered by the local Brit agricultural officer. I can still remember the sad transition of one who came soon after I did to Fort Rosebery. The poor chap, a Scot, as I recall, went from boyish enthusiasm to becoming a regular at the bar after six months or so. The reason was that every scheme he tried failed. The pigeons flew away (or something), tractors were left to rust in the bush and the cassava-fed *Tilapia* fry in the constructed fish ponds were eaten before they could mature and breed. The Bemba were not easy to help.

Aid efforts like this, and the fact that the 'white man' had been the chief source of goodies for several decades, seem to have nurtured a 'dependency culture'. Stephen Biscoe in the *Yorkshire Post* just a few years ago commented "Zambia is poor, underdeveloped, riven with Aids [not known in 1958, although it probably existed, undetected], and being one of the least threatening countries in Africa whose people are famously polite, friendly and easy-going, it is a magnet for the well-meaning..." Then he recounted the following, just one of many examples:

A "*Musungo*" or foreigner, I was told, had devised a way of protecting village women and children from crocodiles when they went to the water's edge to wash and bathe. He lived in an area where such attacks were frequent, and his idea was to create a series of U-shaped croc-proof areas along the river bank consisting of closely placed posts leading down the bank, into the water and back up the bank again. He put his idea to a meeting of village elders, pointing out that the timber could be cut locally, and the stockades constructed by the villagers. The elders discussed the proposal at length among themselves and eventually their spokesman said he had a question. "How much will you pay us to do all this work?'

It is hard to disagree with his conclusion that "Solving problems is not something the local people expect to do – or, indeed, are taught to do." Young Zambian economist Dambisa Moyo made the same point about the unintended consequences of foreign aid in her best-seller, *Dead Aid,* in 2009[8].

A memorable day at Matanda was a visit by some ladies. The party included a French woman, wife of a local store owner, who was visiting someone called Lady Lorna Gore Browne. Lady Lorna, an attractive colorfully dressed woman who appeared to be about 45 years of age, in fact lived nearby, on the banks of the Luapula River in a thatched hut in a Bemba village. We all went for a trip in a native canoe, a hollowed log, on the river and then had some tea on Lady Lorna's little verandah. According to my sister, Lady Lorna called me "Little John" and we got along very well...

I had heard about Sir Stewart Gore Brown and Shiwa Ngandu, but I had no idea what Lady Gore Browne – who was obviously much younger than her ex-husband – was doing in Luapula Province. Her story, which I

---

[8] See also the writings of Scottish-born economist and Nobel prizewinner Angus Deaton: http://www.brookings.edu/blogs/future-development/posts/2015/10/20-angus-deaton-nobel-prize-foreign-aid-radelet

unearthed in detail only much later, is fascinating. Here is an account from her obituary in 2002[9]:

> LADY GORE BROWNE, who has died aged 93, was the spirited chatelaine of Shiwa Ngandu, a manor house deep in the Northern Rhodesian bush. The remarkable story of Shiwa Ngandu, which is vividly told in Christina Lamb's book *The Africa House* (1999), began in 1914, when Lorna Gore Browne's future husband Stewart stumbled upon his "personal paradise".... As a young man, Gore Browne had been in love with Lorna Bosworth Smith, a Dorset girl whose smiling face and serious eyes recalled for him "a Madonna in an Italian painting". Their three year courtship ended when she married, in 1906, Edwin Goldmann, a Polish Jewish surgeon.... In 1927, on a return visit to England, Gore Browne attended the funeral of Lorna Bosworth Smith's grandmother in Dorset. In the church he had to keep himself from staring at a strangely familiar girl in a nearby pew...

That familiar face turned out to be the orphaned daughter of his lost love, Lorna Bosworth Smith. Gore Browne courted daughter Lorna while she was still in boarding school. "One evening, Gore Browne turned up at Sherborne and asked the headmistress if he could take his "fiancée" out to dinner. "We don't have fiancées at Sherborne," said the headmistress. "You do now," he replied."

Eventually, they married even though Sir Stewart was 25 years her senior. Lady Lorna was not a big fan of the Downton Abbey way of life her husband maintained at Shiwa, but she did enjoy the estate and working with the Bemba people. Eventually, disaffected with her elderly spouse and his grand lifestyle – although still friends with him – Lorna left. In 1942 she built this 'country hut' which allowed her to live intermittently amongst the Bemba on the Luapula River. The Gore Brownes divorced in 1950.

Lorna's second daughter and her husband were interviewed at Shiwa by Michael Palin in his "Pole-to-Pole" BBC TV series in 1992. In a tragic postscript, just six months later they were murdered – apparently a robbery, although the real cause is still obscure. Shiwa is now run as a tourist destination by Gore Browne's grandsons.

---

[9]*Daily Telegraph*, February 27, 2002.

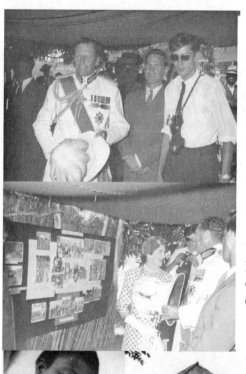

Cool dude JS showing splendidly attired Governor General Lord Dalhousie the Health and Nutrition exhibit in Fort Rosebery ca. 1958.

Lord Dalhousie and wife dutifully admiring H & N exhibit.

Top left: H & N technician Michael Posta getting a blood sample.
Top right: **Main house at Shiwa Ngandu.**
Below: **Portraits of Sir Stewart Gore Brown and Lorna in the house.**

***Trematocranus placodon,* the snail-eating cichlid.**

Here's a more hopeful postscript – about the chronic endemic problem with bilharzia. Fergus and Bent never understood why bilharzia was so much more common at the river site compared to the lake: "There is little doubt that transmission is intimately, if not almost exclusively, associated with the small perennial streams leading to the Luapula river, [which are] frequently and regularly used by the villagers for washing, bathing and other domestic pursuits…" What is so different about these streams compared to the shallow lake waters?

Snails, different species for different varieties of the *Schistosoma* parasite, are one weak link in its imaginative life cycle. *Molluscicide*, killing the snails, is one possible way to attack the disease[10]. Fergus McCollough and several others studied various chemical methods. Not ideal, of course, because of their cost and various ecological side effects. And absence of killer chemicals cannot account for the different incidence of bilharzia in the lake and river sites.

A possible answer appeared just recently[11]: a cichlid fish. The African rift-valley lakes sport the world's most varied set of freshwater cichlid fish, hundreds of species, each adapted to its own special niche. A rather attractive cichlid *Trematocranus placodon*, was once common in Lake Malawi. The incidence of bilharzia was then also relatively low, just as it was in Lake Bangweulu. The reason was that *T. placodon* feeds on the snails that transmit the disease. More *placodon* means fewer snails, hence less bilharzia. When overfishing caused a decline in the Malawi *placodon* population, snail numbers increased and so did bilharzia. Possibly, therefore, it was the relative lack of *placodon* cichlids in the little Luapula

---

[10] McCollough, F. The role of mollusciciding in schistosomiasis control. World Health Organisation, 1992.

[11] Jay R. Stauffer, Jr. and Henry Madsen (2012). Schistosomiasis in Lake Malawi and the Potential Use of Indigenous Fish for Biological Control, in *Schistosomiasis*, Prof. Mohammad Bagher Rokni (Ed.), ISBN: 978-953-307-852-6, InTech, Available from:
http://www.intechopen.com/books/schistosomiasis/schistosomiasis-in-lakemalawi-and-the-potential-use-of-indigenous-fish-for-biological-control

streams that allowed the snails, and thus the bilharzia, to flourish. As far as I know, no one has yet tested this hypothesis in modern Zambia.

In the summer of 1959 it was time to return for my last year at UCL. I flew to London with Eric Dyke, a Rhodesian friend who worked at one of the local stores. The plane was a Vickers Viscount turbo-prop, the first of the type in service. It was a pleasant change from the unpressurized DC3s ('Dakotas') used for internal flights in Rhodesia. Not quite as fast as modern jets, the whole experience was of course much more civilized then than it is now: the windows were bigger and (drum roll!) there was no security check. The plane required five refueling stops, the most memorable at Khartoum, in the middle of the night. I still remember the blast of hot, humid air that came in when the door was opened.

When I returned to England in 1959 I wrote a 4-page article on what I saw as a misperception of the real state of things in Northern Rhodesia. Probably it was intended for Pi, *the UCL student paper. I don't think it was ever published. It gives an idea of how things appeared to a callow 22-year old at that time. Here is an excerpt.*

CENTRAL AFRICA: A MISCONCEPTION

I have recently returned from a two-year stay in a small bush town in Northern Rhodesia. Although I was not able to travel extensively in other parts of the continent, I can at least say that my knowledge of the administration and its effects of the African population is first hand. I can only admire those eminent British politicians who miraculously acquire a thorough knowledge of the country, apparently in all its aspects, within the space of a week or two.

To understand Northern Rhodesia's problems, you need to understand how it is governed. The country is ruled through Provincial and District Commissioners. The Provincial Commissioners are responsible for the provinces into which the country is divided, each vast in extent with populations from 100 to 300 thousand people. Except for the urban provinces of the Copperbelt and the Central Area, the European population is negligible. Each District Commissioner is in direct charge of a fairly small area and their power within this area, both as regards the African and European populations (again, excepting the few urban areas where town management boards exist and the position is rather different), is very great. Consequently the situation made so familiar by Somerset Maugham in his stories of colonial life still appertains. The Provincial Administration, although apparently derided by those not fortunate enough to be in it, is nevertheless much sought after. Although well-disguised, in true British fashion, snobbery is the order of the day and the 'unemployable sons of gentlemen' that comprise the bulk of the administration bask in an adulation that would certainly not be theirs in this country.

It would hardly be fair to the government officers, however, to underestimate the problems with which they have to cope. On the one hand they have the true British Raj contempt for backward races, but on the other they have to toe the Colonial Office line. This line might appear rather surprising to many who have read the reports of Miss Barbara Castle [a very visible Labour MP] and other worthies who have visited the country. It is in fact a very reasonable one. The Colonial Office, far from attempting to retard the progress of Black Africa is very anxious indeed to help the African to find for himself the means to a fuller life.

It may come as a surprise to many in this country to know that every African in Northern Rhodesia. is entitled to free medical attention and free schooling, for example. Of course, these are not, and cannot be, up to European standards, except in the highest grades. Education must be a progressive thing, and until more time has elapsed there are just not enough Africans of an educational standard sufficient for teaching; there is not enough money. It is interesting, however, to note that expenditure on education is second only to Public Works in the Northern Rhodesia budget. This becomes significant when you realise that only the main north-south trunk road and the inter-urban Copperbelt roads are metalled and the Public Works Department is engaged in a heavy programme to improve this situation. Nevertheless, the African High School [Munali] in Lusaka is better equipped than its European counterpart and the teaching standard is higher. This is a result of the system whereby teaching in the European Schools comes under the Federal government and entails many changes of staff from place to place, whereas African education is a Colonial Office concern and continuity is better preserved.

The Administration's problem is a difficult one. On the one hand they have their prejudices as British Gentlemen, and on the other the prevailing policy of African welfare. Luckily for them, however, this problem is made easier for them by an institution known as 'Abolition of Office'. This simply means that should the colony become self-governing and they lose their jobs they would receive a lump sum, amounting to a few thousand pounds, to compensate them for any loss of future career prospects. This is a prospect that is not unattractive to the average officer and enables him to adapt to the 'ridiculous' attitude of the Colonial Office.

As a result of these factors, the Colonial Administration, far from being unfair to the African in fact bends over backwards rather than offend him. This may seem nonsense to many, but I can only report the position as I saw it. As an instance I can cite the response of a District Commissioner to a European complaining of the theft of £180 from one of his stores: "Well, Old Man, you have a hundred stores, the company can afford it." This despite the fact that the thief was known and his Post Office book showed deposits of £50 and £60, against a previous record of 10/-.

It is because of this kind of attitude that the local Europeans seem to many here to be *unreasonably* aggressive. It is not that they are afraid of having a cushy position of superiority threatened, but rather indignation at accusations they feel are the opposite of the truth...

Chapter 5: Roanoke, VA

Saving money was one reason I went to Africa. I succeeded, and when I got to England took delivery of a new tax-free car. I had a choice of reasonably priced cars, between an Austin-Healey Sprite – a tiny, open two-seater sports car – and the new Ford Anglia, a four-seater. Prudence prevailed, and I got the Anglia. Eric Dyke had aspirations to a career in broadcasting. He brought with him a tape recorder with which he hoped to record interviews. It was one of the earliest, an enormous thing that took up much of the back seat of the car.

Car ownership was still a relative rarity in Britain in those days. Motorways didn't exist – the M1 was only begun in 1959 – and rural roads were uncrowded. So, naturally, Eric and I decided to do a tour of Britain. We went as far north as Aberdeen. The weather was wonderful. The lush English countryside was a cool glass of water after arid Africa. I especially remember the Lake District, which was quite stunning in the improbable sunlight. In Keswick we acquired a passenger we met in a cafe. She was a young (we didn't realize quite how young), chubby (we didn't realize why chubby) and very sweet young girl we called "Tiny". I don't think we ever learned her real name. She wanted a lift to London, and we were happy to oblige. Our interest in Tiny was not romantic – she was cute, but more as a puppy than a pin-up. So, even though she shared a room with us a few times, all was innocent.

In those days I seemed to be a destination of choice for very young girls who were looking for a father figure or older brother. I remember two others. One was a 14-year old who worked at Bourne and Hollingsworth in Oxford Street and lived in the dormitory above the store. She appeared at Cedar Road for a party (not given by me... it was all a terrible mistake; things were stolen) and stayed the night in an effort to escape a boring life. But the next day, at my urging, she returned to the fold, or at least to B & H. Another was a very well-dressed young woman, who looked at least 18 or 20 but was in fact also 14, who I met in a jazz club off Piccadilly. She wanted a lift home on the Vespa and described to me how she classified boyfriends into two classes – ones she really liked (who might be romantic or not) and older ones, who spent money on her. I was apparently in the first category.

When Eric and I got back to London, we rented a couple of rooms in Cricklewood while we looked for some kind of flat to lease for the year. Tiny stayed with us, sleeping alternately with Eric and me. On the second day, rather late at night (we were all in bed), there was a knock on the door. It was the police. They had tracked down Tiny who was apparently on the lam from some kind of juvenile-delinquent home in Keswick. It turned out that she was just fourteen, and pregnant (hence the chubby).

She was in Eric's bed that night. Not surprisingly, he was terrified. But Tiny's reputation was apparently well known and the police believed what we told them of our relationship to her. So no charges were brought. Not sure that such common sense would prevail these days. I have often wondered what happened to our little girl.

Eventually Eric and I found a flat – 48a Onslow Road, Richmond. We shared it with my friend Phoebe and possibly Tony Stead (he's in the pictures but not in college by that time). Richmond is quite a way from Bloomsbury but the flat was otherwise a good deal.

Back in Britain, 1959 – the tour, Forth Bridge: *Top:* **JS and Eric Dyke;** *Bottom:* **'Tiny';** *Right:* **JS and Tiny.**

The year passed relatively uneventfully. There were no distracting majors or minors to worry about and I studied more seriously than before. But at first I felt that I had lost at least 20 IQ points from not doing anything in Africa that required serious thought. I did read, mostly nonfiction such as Churchill's *History of the English-Speaking Peoples*. But I did no real intellectual work for two full years. I may have recovered since, but one can't be sure.

I missed a real library in Fort Rosebery. Cricklewood had an excellent branch library just a few minutes' walk from our house. I was a frequent visitor when I was a child, often looking at books well beyond my abilities, like *Ions, electrons and ionizing radiations* by J. G. Crowther (1949), a title I still remember. I'm not sure that modern public libraries would routinely stock books that tough. I read a surprising number of good books, in my three college years – Feigl and Brodbeck's *Readings in the Philosophy of Science* (1953), for example. I can't imagine that this was required by any of my psychology courses. I guess I really was interested in philosophy. I read much Bertrand Russell: in addition to my grammar-school acquisition *History of Western Philosophy*, also *Mysticism and Logic*, *The Analysis of Mind* and especially *Human Knowledge its Scope and its Limits*. Less serious reading was *Marriage and Morals*, and *Satan in the Suburbs and Other Stories*. I dipped into many others from Russell's vast but readable *oeuvre*. I also read some serious novels and biographies: *The autobiography of Benvenuto Cellini,* Italo Svevo's *The Confessions of Zeno,* Dostoevsky's *The Idiot* (never managed *Crime and Punishment*) and many of the Penguin/Pelican series of paperbacks such as Bernard Shaw's *The Black Girl in Search of God* and W. W. Sawyer's *Mathematician's Delight* and a little Roman history: Suetonius' *The Twelve Caesars*. I absorbed several American pop-sociology books: Riesman's *The Lonely Crowd*, C. Wright Mills' *The Power Elite,* William Whyte's *White Collar* and the novel *The Man in the Gray Flannel Suit* (shades of *Mad Men!*) and H. J. Eysenck's hugely successful *Uses and Abuses of Psychology* (which he apparently wrote in just two weeks!) I enjoyed science fiction. Favorites were Olaf Stapledon's *Last and First Men* and stories by Isaac Asimov, Ray Bradbury, Arthur C. Clark and A. E. van Vogt. I also enjoyed the rather strange *My First 2000 Years: The Autobiography of the Wandering Jew*, by Viereck and Eldridge. Alas I lost interest in science fiction as I began to read more actual science.

A book of enormous influence at the time, but now almost forgotten, was Colin Wilson's *The Outsider,* published in 1956. Wilson, an eccentric 24-year-old autodidact, discussed a number of creative people – Dostoyevsky, Kafka, H. G. Wells and several others – who, like all truly creative people (he argued), were isolated from the mass of society. In a sort of reverse logic, this appealed to many of my generation who, feeling isolated, could now be comforted by the thought that they were probably geniuses. In my *Sennet* review I commented sagely "In short, the answer that will serve the common man ["not me!" I hear you cry] can never serve the outsider, he demands something that will neither be comprehensible nor acceptable to the vast majority of mankind." Embarrassing, really. Poor Wilson published many other works, both fiction and nonfiction, on increasingly eccentric topics, but none attained the popularity of *The Outsider*.

Where the impulse for this rather broad reading came from, I don't know. My official course of study at UCL was much narrower than what is usual in American universities. It should have restricted my interests. It did not, and I think the reason is the freedom it allowed – little need to attend lectures, no week-by-week testing. I could have goofed off completely (instead of just partially). I daresay some did (I mention no names). But there must have been something about the culture that directed my interest in relatively worthwhile directions. All I can say is that there is more to a great university than classes and curriculum.

A couple of interesting figures lived in Richmond at that time. One was Bertrand Russell, always a hero of mine, at least as far as his philosophy is concerned. I was not as convinced by his politics. On account of his age – he was 87 – and my own timidity, I made no attempt to try and visit him even though, at 41 Queens Road, his house was within walking distance. A pity: Russell lived on to 98 and was lucid until the end. He would probably have been happy to see a young fan.

48A Onslow Road, Richmond in 1959. Clockwise from top: Tony Stead, Phoebe, Jean Stead, Brian Taylor, Eric Dyke, Ford Anglia.

The other person was Gordon Pask (1928-1996). Pask was an eccentric but brilliant early pioneer in cybernetics, artificial intelligence and programmed instruction. He had a small company, System Research Ltd., on Richmond Hill, a short walk from our flat. I had been struck by his intriguing paper 'Physical analogues to the growth of a concept', in the two-volume collection of papers from a pioneering conference in 1959: *Symposium on the mechanization of thought processes*[1]. Pask was (literally) a theatrical figure, as I later found out. In any event, I screwed

---

[1] London: HM Stationery Office

up my courage and visited him at his company office. He and his colleagues were very hospitable and showed me various gadgets they were working on. But I found Gordon's papers in the main much too difficult and did not follow him far down what turned out to be an almost mystical theoretical road.

The BSc degree was decided by performance on a bunch of exams at the end of the academic year. The exam regimen was much more severe than anything in contemporary America: the outcome of three years of study depended on the results of a single block of tests packed into just a few days at the end. High stress, one might think. Alas, we were not offered anything like Duke University's "Puppies at Perkins, a stress-relief event held during exam week" in which anxious students are comforted by dogs. Comparable cuddly comfort is now offered by many U.S. colleges[2].

I sat six 3-hour written exams and two six-hour practical exams all within a week or so. Despite quite a few fumbles during the practical exams (I had neglected to learn how to use the big mechanical calculators in the lab), I managed to get an upper-second class ("2-1") degree. Not too bad, but I wanted to go on and do research. One usually needed a "first" to be accepted for postgraduate work in England. Lacking a research attachment to any of the UCL faculty, I had no one to say "Well, his degree is mediocre, but he is a bright fellow, good in the lab." So, like many refugees, I turned to America.

My image of America was much affected by American cars. There were a few to be seen in London. They were very large, brightly colored and always in immaculate condition – the epitome of over-the-top luxury. I had earlier been to an auto show and seen a French copy, now defunct, called a Facel Vega. That was also pretty fancy (but the Goddess of French cars is, of course, the Citroen DS19). I remember that a 'mature' American psychology student at UCL had a push-me-pull-you left-hand-drive Studebaker: because of the wrap-around windows on both ends it was not immediately obvious in which direction it was supposed to go. It stood out among the handful of stumpy Austins and Morrises in Gordon Square. A Jewish New Yorker, he had an English name chosen, he said, from the NYC phone book. I saw my first up-close American autos in Africa. There was in Fort Rosebery a flamboyant businessman we all called "Luapula Joe". I never quite knew what he did, but he sure had a great-looking car: a 1959 Chevrolet – a wrap-around, be-chromed, whitewalled and multi-finned monster. That, and the music, was my idea of America.

---

[2] http://usatoday30.usatoday.com/news/education/story/2012-05-13/dogs-stress-relief-on-campus/54921444/1

On the Psychology Department notice board in Gordon Square I found ads for research assistantships from three North American institutions: Kansas State, McMaster University in Canada and Hollins College in Roanoke Virginia. I wrote to all three and got a similar offer from each: $1800 for the academic year. So I had to choose. By any objective standard, I chose poorly. I knew Canada was cold and in any case I was more interested in the U.S. I knew that Kansas was flat and "kinda" far from anyplace I knew about. But Virginia sounded "sorta nice". So I picked Hollins.

It was a few months before an American friend pointed out that Hollins College was (and still is, though it has – inexplicably – changed its name to Hollins *University*[3]) a girls school, at least for undergraduates. The mimeoed information I got from the Hollins psychology people made no mention of this fact. It turned out that Hollins had two co-ed masters' programs: in English and Psychology. The English program is well known for writers such as Lee Smith and Annie Dillard. The Psychology program was pretty good, but the stark differences with the Brit system made the experience a shock at first.

Of course, the rational choice would have been McMaster University. Their experimental psychology program was nationally ranked, larger than Hollins and far beyond it in reputation. But to come to that conclusion just shows what is wrong with a simple notion of rationality, as I will describe later.

So, Hollins it was. In the late summer of 1960 I set off on board a small cargo ship (there were still cargo ships; this was before the era of vast container vessels). 1960 was perhaps the last year when it was cheaper to go to the U.S. by sea than by air – and cargo ships were the cheapest way of all. Ships were soon displaced as the cheap way to get to the U.S. by Icelandic Air which, because of some restrictive trade rule, was forced to stop at Reykjavik on the way. I remember flying on a huge four-engined propjet a few years later. This time I embarked on the U.S. Lines *American Leader*, with a handful of passengers. I shared a cabin with a Chinese guy and made friends with a young English woman going to New York to meet her GI fiancé. I was introduced to root beer, of which I had never heard – and hope never to encounter again.

The ship left from the Port of London, stopped at Le Havre in France and then on to New York. The weather was great and the sea smooth. I was amused by the differences between the three ports. In London, *I'm all right, Jack* prevailed. The stevedores were relaxed, leisurely. Even I, unused to excessive effort or any but British workers, could see that all

---

[3] A reversal of the history of my *alma mater*, which changed its name from the University of London to University College in the 19th century.

was not really as it should be. Le Havre was a bit different. The French dock workers were much more lively. But neither London nor Le Havre could compare with the energy and efficiency of the New York longshoremen (the American term for dockers).

Left: **Loading *American Leader* with a 1960 (Pontiac?) in Le Havre.**
Top right: **Pool of London.** *Below*: **The New York skyline in 1960.**

The British dockers soon had their come-uppance. After damaging industrial action in the early 1970s, container ships, which much reduced the need for dock labor, began their cruise to domination of sea transport. The New York longshoremen also had their problems. In 1969 my parents emigrated to the U.S. They came by ship, via New York, with five crates of their belongings. One was stolen. The NY dockers were efficient as thieves as well as longshoremen, apparently. This also favored the move to containers.

I had an odd experience with the immigration authorities. I was puzzled when the customs guy seemed more interested in whether I was carrying pornographic literature than anything else. D. H. Lawrence's *Lady Chatterley's Lover* was mentioned, possibly because of an ongoing obscenity trial in London in which *Lady Chatterley's Lover* was the star. The government censor accused Penguin, the publishers, of producing pornography. The government lost the case in November that year. My small book collection was a disappointment to the customs man.

New York City was also a bit of a shock. Not the tall buildings, but the cars. Those gleaming monsters I had so admired in London were here in quantity, but mostly dirty and banged up. The cars were a mess.

I stayed the night in a YMCA and then next day took a Greyhound bus (cost: $11) to Roanoke, Virginia. After another few nights in a new "Y" in Roanoke, I rented an apartment and began my studies at Hollins. My second apartment in Roanoke was to the east of Hollins. I drove in every morning facing a sunlit Tinker mountain, covered in beautiful brilliant – much more brilliant than in Britain – autumn (Fall!) colors. I noticed first the big drop in temperature between day and night. And then, later in the season, it got cold, quit a bit colder than Britain. I dredged up a grammar-school-geography memory of the difference between continental and oceanic climates…

Hollins College is nestled in the beautiful Roanoke Valley close to the Blue Ridge Mountains. Founded in 1842 under a different name, it was a residential campus with graceful largely ante-bellum buildings surrounded by green lawns. There was a stable for the horses owned by many of the girls, and several houses on campus for faculty. It had about a thousand undergraduate women and a handful of graduate students. There were only four or so in Psychology, I think. The peace of the valley was later ruined by the construction of Interstate 81. The roar of traffic is now a constant background.

I was supported by Allen Calvin, an entrepreneurially inclined experimental psychologist with a grant from the Encyclopedia Britannica Foundation to do research on programmed instruction. AC went on to become President of Palo Alto University in California and is still active. I helped him with the design and testing of programmed-instruction textbooks in mathematics. I have little memory of the work, because my research interests were elsewhere.

I do remember one book that was much talked-about by the Hollins girls: *The Prophet*, by Kahlil Gibran. I had never heard of it – it made much less of an impact in Britain than in the U.S. I was given a copy and tried dutifully to read it. It is a series of prose-poetry lucubrations on themes ranging from marriage, through crime and punishment to pain and death. A sample:

> [O]f Crime and Punishment. And he answered, saying: It is when your spirit goes wandering upon the wind, That you, alone and unguarded, commit a wrong unto others and therefore unto yourself. And for that wrong committed must you knock and wait a while unheeded at the gate of the blessed.

Like the ocean is your god-self; it remains for ever undefiled. And like the ether it lifts but the winged. Even like the sun is your god-self; It knows not the ways of the mole nor seeks it the holes of the serpent.

But your god-self dwells not alone in your being. Much in you is still man, and much in you is not yet man, but a shapeless pigmy that walks asleep in the mist searching for its own awakening.

Er, yes, I guess... Raised on a literary diet of Bertrand Russell, H. G. Wells and Samuel Johnson, and no poetry after Milton, Coleridge and, maybe, Yeats, I could see nothing in it. The magic of *The Prophet* still eludes me. (Maybe the animated movie version is better – it does feature the wonderful little girl from *Beasts of the Southern Wild*: Quvenzhané Wallis.)

Another shock: no more of UCL's free-wheeling study arrangements at Hollins. We had many regular classes. Class attendance was required – either because the instructor took roll, or because the frequent exams tested for class-specific material. The exams were graded by the teacher.

The obvious conflict of interest when the teacher is also the grader surprised me. The British system was scrupulously designed to eliminate any trace of favoritism or bias. Both the examinee and the examiner were anonymous. Now, in the U.S., students evaluate teachers, both formally in mandated questionnaires and informally on rate-my-professor.com. And administrators use these evaluations to decide on salary and tenure. The incentives for teacher misbehavior, especially by faculty lacking the protection of tenure, are very great. Grade inflation and dumbing-down of course material are just about inevitable. To a Brit, it is rather amazing that U.S. higher education is still as good as it is.

Now, as I write, agitation is roiling even a few elite U.S. campuses, with cries to restrict speech and provide spaces safe from uncomfortable ideas. I wonder how much these new developments owe to a system that grades and may even abuse teachers via the anonymity of social media. There is little doubt that the authority of the teacher has been eroded as the 'needs' of students have been elevated. Is an academic culture that has given us "puppies at Perkins" also responsible, at least in part, for current demands for "safe spaces" and freedom from criticism?

The U.S. system does have one great advantage, though: spontaneity. Because the teacher controls both the syllabus and the content of exams, he (or she) can change both when he feels like it. This makes it much easier to introduce new material or react to some new discovery or interest. The British system, on the other hand, demands a shared syllabus which is very hard to change, and almost impossible to change quickly.

This is not a problem in stable, highly structured subjects like mathematics and the harder sciences, taught at an elementary level. It is a real limitation for the social sciences and some aspects of biology.

I encountered this limitation of the British system many years later, when I was on sabbatical leave at York University in the U.K. I would have been happy to give a few lectures to undergraduates, but it soon became clear that the syllabus and lectures were so tightly linked that no intrusion into the pre-programmed flow was practicable. (Of course, it's possible that my Brit colleagues just thought me a crap lecturer – but they did ask me to give departmental colloquia, so that was probably not the case...)

Possibly the advantages of the U.S. system outweigh its disadvantages. But I found the British system more to my taste because of my self-directed-to-a-fault style of studying, an ability to tackle long examinations without being unduly stressed, and a general bloody-mindedness and resistance to being regimented.

Still the U.S. system was what it was and I had to learn to cope with it. Statistics was the worst. The class met at 8:00 a.m.(!). The teacher had written the text. Worse, much of his teaching consisted of just reading from the text (really!). I survived the course, but only just. I learned much more from a small Pelican book: *Facts from Figures*, by M. J. Moroney.

Although I did not work for him, the faculty member whose interests were closest to mine was Robert Bolles (1928-1994). He was a quiet man with a bit of a limp, from early polio. He did important work, especially on motivation and avoidance and escape behavior in rats. He moved on to the University of Washington in 1966. Bolles was the one who first introduced me to the work of B. F. Skinner, through a book by one of his acolytes, Murray Sidman. I had heard of Skinner of course. But in the books with which I was familiar he had always been discussed in a rather dismissive way. Sidman was different. His groundbreaking *Tactics of Scientific Research* began to open my eyes to a completely different way to study learning in animals, which soon became my interest.

The relations between the sexes at Hollins were also very different from what I was used to in London. There was sexual segregation: girls at Hollins (and neighboring Sweet Briar, now threatening to go bust) and the boys at Washington & Lee, in Lexington a short way up the valley. When the two groups got together, the dynamic was not good. The Washington & Lee boys would descend on the Hollins campus every weekend and their aims were often sinister. It was not marriage they were after. The girls, on the other hand, wanted to get hitched before they graduated. I was not at all comfortable with this rather adversarial arrangement, which was quite different from the more relaxed relations between the sexes I was used to in London.

After a few months, I got to know a rather beautiful and clever girl who everybody called 'DB'. She seemed to be different. She was not a 'real' fee-paying Hollins girl: her widower father was the head of the Art Department. When she was sixteen, she spent a year in Paris with him when he was in charge of 'Hollins Abroad'. She knew Europe, and approved. (While she and her father were away, apparently her Australian mother, destined to die soon of cancer and possibly slightly deranged, who stayed behind with the youngest daughter, sold the house her father had built, leaving him with nowt.) DB was taking an overload of courses,

'DB' – she hated having her picture taken…

and a few 'uppers', trying to graduate quickly, presumably to escape from an unhappy situation.

DB lived with her father and her two sisters in a house on campus that was not, by her account, tranquil. The three girls were not very biddable and her father got mad sometimes, apparently. He always seemed quite amiable to me when I met him, though. And he treated us, and his new grandson, very warmly a few years later. The campus house was always dark – curtains drawn even in daytime. This was a puzzle to an Englishman, who naturally seeks light at every opportunity. The difference of course is the temperature. The house was not air conditioned and although Roanoke is relatively high up, it's still hot compared to, say, Cricklewood.

View from my apartment near Hollins in 1960.

DB had just been dumped by her previous boyfriend, a noisy senior grad student from Michigan with a 1957 Chevrolet convertible (a classic! I think his father owned a GM dealership) who worked on one of the projects. It seemed that he was in search of what the behavioral ecologists call 'resource-holding capacity', i.e. money. He went on to marry a not-so-attractive-but-rich Hollins girl. We all attended the wedding in Towson, Maryland, and put up in a hotel which seemed luxurious to me (heat lamp in the bathroom!). I may even have been best man although Don K and I were far from best buddies. The marriage was short-lived, I believe.

I got one other lesson in Americana during the year at Hollins. I was invited to spend the New Year in Knoxville, Tennessee, with the family of Chris Ruggles, a Hollins friend. The drive down was enjoyable, except for the cool breeze blowing through a crack in the floorboards of my 1952 Ford. An eight-year-old car was in those days regarded as well past its sell-by date, and the Ford was. But some things have improved. My current car is almost that old, but, as far as I can tell, like new in all essential respects.

My new friends were incredibly hospitable, even lending me one of their cars, a 1955 BelAir convertible, to tour around the neighboring mountains. But the thing that struck me most forcibly was American football. A team called the Green Bay Packers were playing another team with a less memorable name. I tried to watch, encouraged by a box on the screen announcing just a few minutes were left to play. It was not to be. They were kidding about the time. Used to real football (i.e., soccer), I couldn't believe the constant *sportus interruptus* that stretched out the time from a few minutes to half an hour.

I now realize that these interruptions are not an accident. They represent a symbiosis between the rules of the sport and the need for advertising. A sport lacking the *interruptus* feature would pose real problems for live commercial TV.

In the New Year, I had to think about what to do next. To do research, I needed a PhD. The days when many eminent scientists had nothing beyond a bachelor's degree were long gone, even in Britain. Hollins only gave a master's degree. So I needed to apply to graduate school somewhere. Two more tests were required: the Graduate Record Examination and the Miller Analogies Test (MAT). I took them at the University of Virginia, in Charlottesville. Both were basically IQ tests, and I did well, despite losing some time fumbling during the MAT. So my prospects for admission to a good graduate school looked fair. I applied to three: Harvard, Indiana and Stanford. I soon heard from Stanford: no dice. (I got in to Indiana too, but they responded late, assuming I must have already been accepted to Harvard or Stanford.) Then I was admitted to Harvard, but with no money. Allen Calvin called someone he knew at Harvard and I got an assistantship from Jacob Beck, a perception expert. It was not enough to live on as well as pay the fees, but if I could supplement it in some way…

In the Spring, I managed to drive up to Cambridge in an increasingly unreliable car and stayed at the home of Bea Anne Greene, a kind Hollins girl and her family on the lake at Wellesley College. I wanted to see what Harvard looked like. The Psychology Department, which was really just experimental psychology, was in the basement of Memorial Hall. The rest of psychology was in another department, Social Relations. The Center for Cognitive Studies, a new development headed by George Miller and Jerome Bruner, was also elsewhere.

Memorial Hall is a wonderful over-the-top red-brick Victorian gothic structure, a sort of Yankee St. Pancras Station. It lacked its spire – lost in a fire a few years earlier and not restored until 1996. It had an enormous basement of more than 100 rooms. At one end was the pigeon lab and the office of B. F. Skinner, already famous as a radical behaviorist and critic of much of psychology. More important, he was the inventor of a powerful new experimental method for what he called *operant conditioning,* a way of studying animal learning with individual animals rather than groups. At the other end of the basement was the anechoic empire of S. S. "Smitty" Stevens, king of psychoacoustics and the major political force in the department at that time. In the front office lived the formidable and efficient Geraldine "Didi" Stone, Smitty's assistant, girlfriend and soon-to-be wife. In the middle was the Chairman Edwin Newman. An inhabitant with largely nocturnal habits was a Hungarian émigré, Georg von Békésy, an expert on the psychobiology of hearing and about to win a Nobel Prize. He was a lecturer, basically an adjunct appointment, with limited teaching duties. A few young professors and "postdocs" made up what was a very small department. Most notable from my point view were assistant professor R. J. Herrnstein, lecturer George Reynolds and postdoctoral fellow Charles Catania.

Dick Herrnstein showed me around. I was much impressed by the wonderful experimental methods on display – visible evidence that it was possible to study the behavior of individual animals in a direct and powerful way – no statistics, no group data. I had always hated the idea of studying animals in groups; statistics seemed to miss the point. But until I saw the operant lab, I had no idea that an alternative existed which permitted precise experimental control and quantitative recording of individuals.

One lab that made an impact belonged to an Australian graduate student, Peter van Sommers. Peter was a genius with equipment and a beautiful experimenter. He was studying the effect of oxygen reinforcement on the behavior of goldfish. I was most impressed to see a little fish swimming upstream in a Plexiglas tube and nosing a plastic disk that delivered pulses of oxygenated water.

Peter is the son of an artist; his father had been an official war artist during WW2. Peter was always interested in photography. When he left Harvard to travel the world he showed me an ingenious gadget with a right-angle prism that allowed him to take un-posed pictures of people while he seemed to be looking away from them. His interest in art increased and in later years he became a professor of design in Sydney[4].

The basement of Memorial Hall, was not elegant – almost every room had a ceiling full of heating pipes serving the grand halls above. But it was wonderfully efficient. With many rooms, all on one level, for half a dozen faculty, all was accessible and there was no shortage of space. William James Hall, the boxy new 1960s-style[5] 15-storey building to which both Social Relations and Psychology were moved the year after I graduated, is different. It is a triumph of (dull) form over (mal)function. The Mem. Hall basement was all on one floor. It was impossible not to bump into your colleagues in the course of the day. Smitty Stevens psychophysics types constantly interacted with the operant people. And all shared secretarial, workshop and drafting facilities. In contrast, WJH is perfectly designed to prevent accidental interaction between people in different specialties. Each rather small floor is devoted to one like-minded group and getting from one floor to the next requires either the elevator or going through four fire doors and up or down stairs. The perils of intellectual miscegenation have been designed away.

I was delighted to be going to Harvard. It would never have happened had I decided a year earlier in England to go to McMaster rather than Hollins. Without the Hollins experience I would not have got into Harvard, nor would I have been able to cope even if admitted. My 'irrational' choice of Hollins turned out in the end to be best. So, in the late summer of 1961 DB and I (for we were now together) rented a U-Haul trailer into which we packed everything, and headed for Cambridge MA. We had just three pieces of homemade furniture: two nice little coffee tables and a large rather eccentric table lamp (my fault) that we had made together in the Hollins Art Department's excellent woodworking shop.

---

[4] See his wonderful art at http://www.petervansommers.com/

[5] It was designed by Minoru Yamasaki, who was also the architect for the ill-fated World Trade Center.

Chapter 6: Cambridge, MA

The U-Haul was clamped to a ball on the back bumper of our car. On the way to Cambridge, somewhere in the middle of Washington, DC, the ball dropped off and was lost. The holding nut must have come unscrewed and fallen off. We didn't lose the trailer – it just dragged by the safety chains. But we were stuck. Amazingly, an angel appeared almost at once. A man in a passing pickup truck saw our plight. He stopped and offered to give us a new ball. I detached the trailer, drove behind him to his house and got the ball. I returned and we attached the ball and we went on our way. The chance that someone who knew the problem, had the solution, and was willing to provide it, is surely very small. A tribute to Lady Luck – and to the kindness of Americans. I'm pretty sure we would not have been so fortunate in London.

In Cambridge we stayed in a B & B for a few days until we could get an apartment. The main thing I remember was the mosquitoes. We were bitten to death that first night. Based on similar experiences in Ontario a few years later and comparison with Virginia and North Carolina, I can conclude (this is definitive): the North is worse for mosquitoes than the South. After a couple of days, we rented a tiny rooftop apartment in Greenough Avenue then, for the last two years, moved to pleasant Harvard Graduate Student Housing in Shaler Lane, near the river. I set up a little aquarium, with an Oscar cichlid. He (I think it was a he, but male and female Oscars look the same) seemed to be quite smart: he changed color and acted alarmed when someone other than the two of us looked in; he disliked a particular check shirt. He grew from a couple of inches to 8 or 9 after two years.

We eventually got a dog. My parents arranged for their beloved cocker spaniel Blondie to be flown from Rhodesia to Boston when they returned to England on leave. The six-month quarantine regulations then in effect made it impossible to take the dog with them.

But our first pet was a cat. She had been in a study on maternal behavior run by Rhoda K, one of the graduate students. Unfortunately the cat miscarried while giving birth. She survived, alive but neutered. No further use to the study, she had two possible futures: either euthanasia or adoption. We were happy to adopt her.

How were *we* to live? That was the next question. It was solved by DB getting a job – at Massachusetts General Hospital as an assistant to Dr. James Anliker, a one-time Skinner student. With her salary and my assistantship, we could weather the first year. After that, I got Harvard credit for my Hollins courses so outgoings were much reduced.

I was surprised to learn later that Dick Herrnstein, Rhoda's supervisor, and subsequently mine also, was very critical of her decision to give the

103

cat to us rather than kill her. Reprieve was a violation of the rules, apparently. But the rules, designed for the safe disposal of damaged animals, made no sense in this case. Our cat was perfectly normal, had not been exposed to drugs or pathogens nor had bits of her brain removed. This was an early hint of the steady transmogrification of the laboratory-animal-welfare regulatory apparatus over the years. It began by being a bit rigid and a little stupid. As the bureaucracy expanded and grew increasingly insulated from legislative oversight, the rules multiplied. They began to reflect the needs and incentives of the American Association of Animal Laboratory Care (AAALAC[1]) and political pressures from animal-rights activists, rather than the divers realities of research – or even the real needs of the animals. Animal-care compliance has become an expensive and often infuriating obstacle to laboratory research with animals. The impact is worst on non-invasive research, which is less costly, but also less-well-funded, than 'wet' neuroscience. Of course, it is also less harmful to animals, but the same rules apply, mindlessly, to all.

In what must be a record for casual matrimony, DB and I soon got married. DB wanted to and I saw no objection. We sought out the oddest Justice of the Peace we could find, a Chinese guy, Tehi Hsieh by name. Witnesses were hangers-on in his office. I noticed a few years later that the marriage certificate gives an expiration date for Tehi's license that is in fact earlier than our marriage date. Was this an omen?

We had no grasp then of the obvious fact that marriage is not just an individual thing. It is introducing a new person into two families. They should at least be consulted. I can perhaps be forgiven, since my family was thousands of miles away, in Africa. But we basically eloped, so DB's father wasn't in on it either. We might have understood better what we were getting into if our families had been permitted a comment or two. DB and I got along pretty well for two years, under rather difficult circumstances because of my workload. But then we had a big fight. I can't remember what about, but I do remember that she ran off into the street and I had to go looking for her. Things settled down after that, but...

Getting a doctorate in America is more complicated than in Britain. In the U.K., it is a purely research degree. You just do the experiments and produce a weighty work of scholarship. If that passes muster, you have the degree. But in the U.S., there are courses and exams – fewer at Harvard than many other places, but definitely non-zero. In Mem Hall the first big

---

[1] AAALAC, now rebranded as the Association for Assessment and Accreditation of Laboratory Animal Care International. The acronym must be preserved, apparently!

hurdle was the Pro-Seminar (ProSem), a two-semester course taken by all the half-dozen or so first-year students.

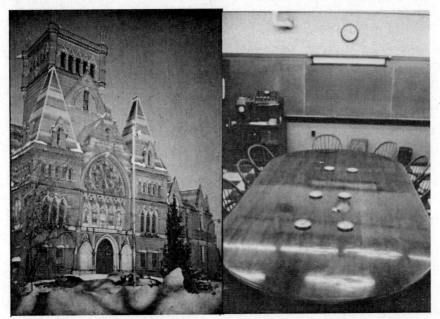

Memorial Hall, Harvard, in winter 1962, *sans* spire, which had burned down in 1956 (restored in 1999).

'Pigeon staff' seminar room in the Memorial Hall basement.

First impressions tend to be indelible. My first impression was of Smitty Stevens running the ProSem with an unforgiving hand for most of the first semester. We got a thorough indoctrination in psychophysics, the science of sensation – especially psychoacoustics, Smitty's specialty. We learned of Smitty's useful distinction between different types of measurement: nominal, ordinal, interval and ratio.

Smitty Stevens was a Mormon and grew up with his polygamous grandfather in Utah. One proof I have of his Mormon origins is an abandoned maple wood piano stool, labeled LDS-CHURCH, discarded by Smitty. It was broken; I took it and fixed it, DB re-finished it and I have it still.

Speaking of Mormons, the Bible for the ProSem was Stevens' 1951 *Handbook of Experimental Psychology,* a 1400-page door-stopper. I read it dutifully for my presentation on psychoacoustics. Not good enough! I failed to mention the big news in 1961, which was *Steven's Power Law*, a quantitative relation between the level of a sensation – a number produced by the subject – and its physical intensity. A tone of 90 decibels (dB), say, might be given a value of 20, by a subject, 100 dB might be called 27, and so on. Stevens and his students found a simple relation between these numbers and the corresponding decibel values: the judgment is proportional to the physical value raised to a power – less than one for light and sound intensity, more than one for electric shock. Moreover, you can predict how people will adjust the intensity of one stimulus, a sound say, to match the intensity of another, a light, from their separately obtained power functions. This is called cross-modal matching and was taken as a great vindication for the power law.

The presumption that experimental psychology is all about finding general laws pervaded Memorial Hall. The idea has a venerable history, beginning with 19th century German scientists Weber and Fechner and their psychophysical law. E. H. Weber found by experiment that for many intensive dimensions – weight, loudness, brightness – an individual's ability to detect a change in level is proportional to the base level. A Just-Noticeable Difference (JND) of 1 lb from a base of 10 lb implies a JND of 2 lb from a base of 20 lb. G. T. Fechner then deduced that if all JNDs are subjectively equal, the strength of sensation is proportional to the logarithm of the physical stimulus value. Stevens' power law was considered the next great advance on Weber-Fechner.

The Power Law

So, is sensation related to stimulus by a log function or a power function? Who is right: Stevens or Fechner? Well, both apparently. In 1963 Donald MacKay[2], a brilliant British cybernetics type, showed that you have to take the psychophysics of the number scale into account. His idea was that sensation is indeed logarithmically related to the physical stimulus, whether it be a light, a sound – or a number. With a log transform ($\rightarrow$) at both ends: stimulus→sensation→number, stimulus and number will be related by a power function.

In a theoretical paper years later[3], I was able to explain some novel empirical results with this idea. In one very elegant experiment, published just after I left Harvard, Stevens and a colleague had three subjects adjust the luminance (intensity) of a matching field (seen only by the right eye) so that it matched the apparent brightness of a target field seen by the left eye. A small, intense glare source was also visible to the left eye. The experimenters adjusted the brightness of the target field and thus obtained a series of matches. They varied the separation between glare source and

---

[2] MacKay DM. Psychophysics of perceived intensity: a theoretical basis for Fechner's and Stevens' laws. *Science*. 1963;139:1213–16.

[3] Staddon, J. E. R. (1978). Theory of behavioral power functions. *Psychological Review, 85*, 305-320. http://hdl.handle.net/10161/6003

target and got different power functions for each glare level. The relation between target luminance and matching field luminance was a power function whose slope and intercept (in log-log coordinates) depended on the visual angle between glare source and target: the more the glare, the steeper the function – presumably because the glare reduced the subjects' ability to detect small changes (i.e. his differential sensitivity). The closer the glare, the larger the target-luminance JND.

Figure 6.1

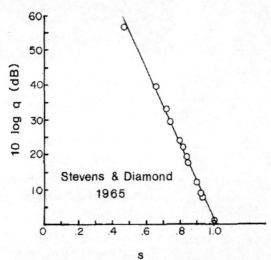

Figure 4. The relation between the slopes (*s*) and intercepts (10 log *q*) of power functions fitted by S. S. Stevens and Diamond (1965) to matching data from comparisons of the perceived brightness of a match field and a target field inhibited by an adjacent glare source.

MacKay's idea of an output as well as an input log function implies four parameters[4], two for each, a scale parameter and what might be called a 'sensitivity' parameter. If MacKay is correct, the two parameters of the overall stimulus-response power function incorporate these four.

As a test of the log-log idea, I proposed that glare in Smitty's experiment has an effect only on the sensitivity parameter of the input log function, not on the scale parameter or either parameter of the output. The math then shows that the power functions that Stevens found at different glare levels should all converge on a point – which implies that the slopes and intercepts of the functions should all fall on a straight line. The data

---

[4] The Weber-Fechner Law is $y = a\log x + K$, where y is sensation, x is the level of the physical stimulus and K is a scale parameter; a is the sensitivity parameter.

match this prediction perfectly (Figure 6.1). In other words, Smitty's own data show that the power function derives from the more fundamental logarithmic input and output functions.

Top left: **Mrs. Antoinette C. Papp, Queen of the Memorial Hall pigeon lab.**
Middle: **Weighing a pigeon after each experimental session.**
Bottom left: **The pigeon room (in violation of current AAALAC regulations).**
Top right: **Shaler Lane: Harvard Graduate Student Housing, in 1962, with our blue 1952 Ford. Much kiddie stuff is in evidence.**

Smitty died in 1973, before my paper was published, but I don't think MacKay's work convinced him that his power law derives from Weber-Fechner. My paper in fact made little impact even though it appeared in psychology's most important theoretical journal. Possibly because I had no publication record in the area of psychophysics. And possibly also because the paper began with a discussion of power functions in operant conditioning and psychopharmacology. The Stevens discussion, with the best data, conclusive proof and speaking to a live issue in psychophysics, came at the end. MacKay's point was reiterated some 25 years later – alas with no acknowledgement of his work or mine[5].

After Smitty, Edwin G. Boring, the founder of psychology as a separate department at Harvard and respected historian of psychology, took the ProSem for three weeks. EG was long retired at that time and

---

[5] Dehaene, S. The neural basis of the Weber–Fechner law: a logarithmic mental number line. *TRENDS in Cognitive Sciences* Vol.7 No.4 April 2003.

regaled us with stories and quotes from the German origins of experimental psychology. All I can remember is his awful German accent.

More memorable was B. F. Skinner, who also only got three weeks. He was just 57 years old at that time but no longer personally involved in laboratory research. He had spent the past several years on larger topics, like teaching machines, the social role of radical behaviorism and the book he thought his most important: *Verbal Behavior.* But he turned up pretty regularly for the weekly 'pigeon staff' meetings when his students (now faculty or postdocs) in Mem Hall and from the Harvard Medical School (Bill Morse, Roger Kelleher, Peter Dews, Nate Azrin, Murray Sidman…), and their graduate students, would present data.

I was puzzled by three things: Skinner's insistence that we all learn his terminology, even for concepts like, say, instrumental behavior or secondary reinforcement (now to be called operant behavior and conditioned reinforcement) that already had perfectly good names; his relative lack of interest in new experimental findings; and, most surprising of all to me, his total lack of interest in theoretical explanation. The theoretical problem came up again when I took another class with him later.

I have written at length[6] about my view of Skinner's ideas and methods, so I'll just summarize here. Skinner's idea of the *operant* resembles the folk psychology idea of a *habit.* Nevertheless, treating it as a unit of behavior analogous to the chemist's atom has been fruitful. I love Skinner's experimental methods and the rich lode of new phenomena – schedules of reinforcement – to which they have given rise. But I'm embarrassed by much of the theory and philosophy that accompanies them. I say "embarrassed" because I still cannot understand how a man as brilliant as Fred Skinner could seriously propose and appear to believe an epistemology that makes little scientific or even logical sense. His genius failed to lead him to a better view of theory. But his rhetorical talent and accidents of cultural coincidence – post-war America was ready for simplicity and the promise of improvement through science – gave his ideas enormous influence. Although it has almost disappeared from mainstream experimental psychology, radical behaviorism with its private language and specialized techniques is still the guiding philosophy for thousands of behavior analysts.

The theoretical issue came up again in a seminar with Skinner the next year. He handed out a 'purple-paper' (mimeographed) copy, dated October 23, 1963, of some notes on the concept of the stimulus. It

---

[6] Especially in Staddon, John (2014) *The New Behaviorism* (2nd edition) Philadelphia, PA: Psychology Press.

summarized his important article 'The generic nature of the concepts of stimulus and response' which restated John Dewey's view that each is necessarily defined in terms of the other. No problem; but what happened in between stimulus and response – to the organism – was unstated. An animal is presumably changed by the stimuli it has experienced and the responses it has made. But Skinner was silent on what these changes might be or how they could be represented. Instead of any reference to some kind internal state, all was to be explained by history: the organism's history of reinforcement.

I had many problems with this. I agreed that history is key. But history surely includes more than reinforcement. I am changed by visiting a new place, even though nothing reinforces me for looking around and remembering where I have been. Unless *all* behavior is attributed to reinforcement. But a theory which explains everything explains nothing, so that can't be right. To say that present behavior is determined by past history – to say no more than that – is just to say that behavior is determined. Causality is affirmed, but the possibility of further explanation is ignored.

I sympathized with Skinner's reluctance to get entangled in the mysteries of neurophysiology. No doubt the brain is changed by experience. Any attempt to get at this – as, for example, Ivan Pavlov tried to do – through purely behavioral experiments is surely doomed to fail. But physiology is not the only option.

The first elective course I signed up for in Cambridge was not at Harvard but at MIT. It was John McCarthy's course in artificial intelligence. McCarthy was a mathematician, the inventor of the LISP programming language, so the course was more about predicate calculus than technology. McCarthy was proud of his programming skills, though. At a time when programs were submitted on punch cards to a central computer that would only disgorge the results (usually in my case, errors) the next day, McCarthy claimed never to make a coding mistake. I made many mistakes and learned a little LISP programming. But mainly I learned to think clearly about historical systems. A big help a few years later was a book by McCarthy's colleague Marvin Minsky (who usually taught the course): *Computation: Finite and infinite machines* (1967).

Internal states

Emboldened by this experience, I prepared four pages of my own purple paper notes and submitted them for the Skinner's seminar. I extended Skinner's idea of the generic nature of stimulus and response to a third construct: *internal state*. Skinner pointed out that stimulus and response should be defined in terms of one another. The set of events that cause a set of behaviors represents an 'operant'. The operant is created by repeated exposure to a particular reinforcement schedule. Any member of the stimulus set can produce any member of the response set. Hence the 'generic' property. That was the limit of Skinner's theorizing. I argued that the same logic should apply to an organism's *history*. A set of histories that are equivalent in the sense that the organism responds in the

<u>A simple-minded formalism for talking about behavior</u>

An organism both acts upon the environment and is acted upon by it. What is the simplest formalism that will take account of this fact? The following is offered as a possible (by no means original) candidate, in the hope that it constitutes a language in which may be expressed all and only meaningful (testable) statements about behavior. Since it is probably not adequate, its real purpose must be to encompass its own destruction by yielding something more satisfactory.

Consider the organism as a black box, thus:

$$I \xrightarrow{\quad} \boxed{S}^{M} \xrightarrow{\quad} O$$

I = set of all <u>inputs</u> to the machine $I = \langle i_1, \dots, i_n \rangle$
S = set of all <u>states</u> of the machine $S = \langle s_1, \dots, s_m \rangle$
O = set of all <u>outputs</u> of the machine $O = \langle o_1, \dots, o_k \rangle$

These three concepts are defined as follows:
1. Any arbitrarily defined event (not necessarily a physical property, but conceivably a relation etc. . . .) is defined as an input $(i_i \in I)$ if and only if it <u>either</u> produces a correlated output $(o_j \in O)$ from the box <u>or</u> changes the internal state of the box. $([s_i \in S] \longrightarrow [s_i \in S])$ in some way, or both.
2. A state $(s_i \in S)$ of the machine or organism is defined as a relation between some input $(i_i \in I)$ and some output $(o_i \in O)$ i.e. the relation $i_i \rightarrow o_i$ meaning "the application of input i, yields o_i as output" defines some state, S_i, of the machine. In physical terms no explicit indentification of S is necessary and indeed it is quite arbitrary (from the formal point of view, if not from the point of view of empirical convenience) which parts of the machine we label S and which O.
3. Any identifiable event associated with the machine may be defined as output, provided only that consistency is maintained. From a practical point of view at least two criteria for defining output suggest themselves:
1. It seems intuitively obvious that as much of the machine as possible should be defined as output, if only because by so doing we maximize our information about the machine.
2. Both output and input are defined so that the simplest possible description of any given machine is obtained, i.e. a minimum number of states is required to account for the behavior of the machine.
[footnote: Although a state is defined as a relation between some i_k and its output, o_k, there is no reason to restrict each state to only one input and output. A state may describe many input-output relations:
$$\begin{pmatrix} i_k \rightarrow o_k \\ i_n \rightarrow o_n \end{pmatrix}$$
provided only that an input which changes that state changes all these relations in the same way.]

Quite obviously there will often be many possible true descriptions of a given machine, since a) our knowledge of the internal structure of the machine will always be incomplete and b) if the machine is at all complicated it will usually be impossible to exhaust all the possible input-output relations.

Page 1 of a 4-page JS handout in Skinner's class at Harvard in 1962. My initial attempt to offer an alternative to stimulus-response behaviorism.

same way after any member is what I called a 'state'. The idea is a pretty straightforward extension of Skinner's generic stimulus and response. But 'state' still gets criticism – largely because of the word, I think. Some critics assume that 'state' or 'internal state' must somehow refer to some physical entity inside the organism – rather than simply one or more inferred variables.

The 'state' idea is simple and, as far as I could see, logically impeccable. Suppose first that the effect of any new stimulus depends upon the organism's history until that point. Suppose further that we have an indefinite number of replica organisms, all identical and starting in the same state. We expose organism one to history one; organism two to history two and so one for an indefinite number of organisms and histories. Then we expose all to our new stimulus, call it stimulus zero. We can then divide all our replica organisms into two groups. One group, call it Group A gives response zero (say) to stimulus zero; all the rest give a different response. Since we have an infinite supply of organisms, we can look at another set of Group A and try another stimulus. Do they all give the same response to this new stimulus? If not, we can further subdivide Group A. So now we have organism that response zero to stimulus zero, one to stimulus 1. By continuing this process, we can finally identify all the sets of histories that are equivalent in terms of the organism's future behavior. These equivalent histories are the *states* of our hypothetical organism. A given state is defined by the fact that the organism behaves afterwards always in the same way.

This is a 'thought experiment', impossible to do in practice. But it makes two points. The first is that it is perfectly possible to discuss the behavior of a historical system like a pigeon – or a person – in terms of states which are not physiological, but simply summarize the effects of histories that are equivalent in terms of the system's potential future behavior. After a history lacking in food, after exercising, or following an appropriate appetite-inducing injection, a pigeon is going to be hungry, for example. Nothing is lost by calling the bird "hungry". No concession to mentalism or introspection is implied by this label. It is agnostic on exactly how, in a physiological sense, hunger actually works. "Hunger" just defines a state in which, for example, food is reinforcing; food-seeking behavior occurs, often spontaneously; known sources of food are re-visited and so on.

Second, these states need not, perhaps should not, refer to concepts from human consciousness, from introspection, especially if we are speaking of pigeons not people. Ideally they should refer to things that can be measured in some way.

I cannot remember Skinner's reaction to what I modestly thought to be a brilliant, albeit obvious, proposal. Probably there was none.

Habituation – and eating in meals

Here is an example of how this approach can summarize a set of histories. Habituation is a form of learning in which the effect of a neutral stimulus gets smaller each time it is presented. A loud noise will cause a dog to turn his head looking for the source. But after the noise is repeated a couple of times, he is likely to ignore it.

Habituation has been studied in a wide range of organisms. *Caenorhabditis elegans*, a small nematode worm beloved of molecular biologists has yielded a particularly nice set of data. Swimming in a Petri dish, a number of these worms will turn in response to a tap on the dish. With each tap, fewer turn: they habituate. This behavior can be summarized by a simple model in which each tap charges up a memory that inhibits response to the next tap. The memory is like a leaky bucket (the leak represents forgetting): each tap adds a cup of water which increases the water level in the bucket. If the taps are spaced too far apart, the bucket empties (forgets) in between taps and there is no habituation: the response continues to occur. If the taps are frequent, the bucket fills, the response weakens and may finally cease. The system (animal) habituates. The habituation process in this model is defined by just a single number: the leak rate (termed a *time constant* in the formal model). The bigger the hole, the faster the time constant, the quicker the system recovers after stimulation.

Notice that the memory here is entirely hypothetical. No physical store is implied. The idea of the model is simply to summarize a wide range of data in the simplest possible way. This is called *parsimony* and I take it to be a fundamental principle of science. Skinner agreed with the general idea of *induction*, of theory as "a formal representation of the data reduced to a minimal number of terms"[7]. But he never took the next step, which is to test the "formal representation" *de*ductively. His aim was elsewhere. He wanted not so much to explain behavior as to control it. Once he found an effective means of controlling the behavior of pigeons, his interest in their behavior was essentially at an end.

The leaky-bucket memory model for habituation is now an old idea, first suggested by the Russian scientist Evgeny Sokolov in 1960. It explains the effects of stimulus spacing on the rate of habituation: stimuli at short intervals yield rapid habituation, at long intervals slow or no habituation.

But Sokolov's idea does not explain a paradoxical effect called *rate-sensitivity*. When taps are spaced far apart the level of habituation is reduced – not so many worms turn. The leaky bucket can handle that: less frequent stimuli (cups of water) means that there is more time for the water to leak out, hence the water level is lower; hence the response is less inhibited – there is less habituation. But after a series of widely spaced taps, habituation persists longer (shown by tests at different delays) than after training with frequent taps. In other words, even though the level of

---

[7] *Science and human behavior*, Skinner, 1950, p. 69

habituation is greatest when taps are frequent, it can persist for longer when they are less frequent.

This effect of spacing is very general, showing up even in complex learning. Despite its obvious adaptive significance, rate sensitivity has been largely ignored by habituation researchers. Many years after Sokolov and my time in Memorial Hall, Jennifer Higa and I were able to show that rate sensitivity requires the equivalent of at least two leaky buckets[8]. The system is defined by at least two numbers, two time constants, not one. Our little model summarizes the results of many possible histories, experiments with an indefinite number of possible trial spacings.

Figure 6.2 *Top* **Pattern of eating of 6 rats over a 24-hour period.** *Bottom* **Pattern of eating generated by 6 model rats.**

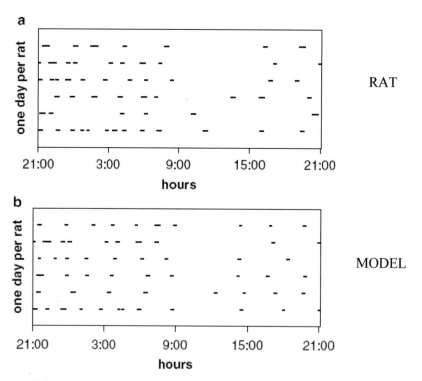

This simple approach recently allowed Silvano Zanutto, a colleague from Buenos Aires, and me to explain some well-known facts about

[8] Staddon, J. E. R., & Higa, J. J. (1996) Multiple time scales in simple habituation. *Psychological Review, 103,* 720-733.

eating behavior[9]. Animals do not feed at random. Even grazers, like cattle and Canada Geese, who must ingest low-energy-density food for many hours each day, tend to prefer to eat at certain times. More interesting, though, are the eating patterns of omnivores like rats, cats, dogs – and humans. Their food usually has more energy than grass, so they need not spend all their time eating. Omnivores, eat in *meals*, relatively short bouts of concentrated eating that are roughly periodic and separated by intervals when the animal does something else. We wondered if there might not be some simple principle behind the pattern of meals and the way that the pattern changes if feeding is restricted in various ways.

Research on the physiology of eating has long assumed that eating must increment some kind of *satiety signal* that eventually terminates each meal. Physiologists have searched diligently for the source of this signal. Is it blood sugar level? Body fat? Some circulating hormone? It seems that satiation, like animals' sense of time (which I get to in a moment), depends on many physiological factors. There seems to be no single 'internal clock' and there is no single physiological satiety signal (although Silvano was able to narrow down the brain areas that seem to be involved).

Whatever the details, the regularity of eating in meals under normal conditions suggests that the satiety signal must have some relatively straightforward dynamic properties. Silvano and I began with the notion that there is a threshold, θ, so that when the satiety signal (call it V) is less than θ the rat eats, and when V is greater than the threshold, he quits.

The key question was: what determines the value of V? Our answer was a system very like the memory system that underlies rate-sensitive habituation: a chain of 'leaky buckets'. As the animal eats, the buckets are filled; but with time they empty out. The result is satiety signal (amount of water in the buckets) that rises as the meal progresses, finally crossing the 'stop' threshold. After the meal ends, V continues to rise for a while but then slowly declines until it again crosses the threshold and initiates another meal: the satiety 'stop' signal is thus delayed after eating onset. The key feature of this little system is *delay*: V rises to a maximum as the animal eats and declines to zero some time after he ceases.

Three buckets (a two- or three-parameter system) are sufficient to generate a satiety signal that duplicates the usual pattern of eating in meals (Figure 6.2 – the gap in the record after 9:00 represents a slightly higher threshold during the nighttime). If, when the animal starts to eat, feeding is blocked for a period longer than the typical inter-meal interval (but less

---

[9] Zanutto, B. S., Staddon, J. E. R. (2007) Bang-Bang Control of Feeding: Role of hypothalamic and satiety signals. *PLoS Computational Biology,* 3(5): e97 doi:10.1371/journal.pcbi.0030097

than a day) then, unsurprisingly, the animal compensates when allowed to eat again. Not so obvious, though, is that all the compensation takes place in an extra-long first post-interruption meal, the so-called 'first meal effect'. Longer interruptions, of a day or more, produce somewhat more complex effects which are also matched by the model. The model also predicts the way that the animal adapts when he must work (on a kind of ratio schedule) to get food.

Because the satiety signal continues to rise even after eating ceases, if a meal is interrupted in the middle (say) and the animal is allowed to resume after a delay he should begin to eat again if the interruption is relatively long, but not, perhaps, if it is relatively short. No one seems to have tried this test.

Skinner made only one attempt at formal theory. His model of the *reflex reserve*, proposed very early on, is similar in many ways to the Sokolov model. But he soon abandoned real theoretical work. Someone who did not was R. Duncan Luce, then the reigning monarch of mathematical psychology. He also took a week or so of the ProSem. Much of what he said was beyond my expertise, but I survived the exam by learning one or two things pretty well (though I couldn't say now just what they were).

Sequential effects

I did have an interesting contact with Luce's work some years later, in collaboration with a colleague and a graduate student at Duke University. In *absolute judgment* experiments, a subject is presented, with stimuli, one after another – loudnesses, say. He is asked to identify which one of (say) ten different loudnesses just occurred (this is called *category scaling*). If the loudnesses are not too far apart, subjects will make errors, overestimating some stimuli (calling loudness no. 5 "7", say) and underestimating others. Greg Lockhead, in one of many experiments in which feedback was given after each trial, found a very simple sequential relationship between the size and direction of the average error and the stimuli on previous trials. For example, suppose you have a set of ten stimuli (loudnesses). People tend to underestimate after a quiet stimulus but overestimate after a loud one (this is called *assimilation*). But on the next couple of trials they tend to make mistakes in the opposite direction: underestimating after loud and overestimating after quiet stimuli (this is called *contrast* – Figure 6.3)

When there is no feedback, people show assimilation only. Possibly confused by this difference, Duncan Luce and his colleagues made an elementary mistake in describing the underlying process. It is very easy to show that assimilation by itself requires only an effect of Trial N on the succeeding Trial N+1. But absolute judgment with feedback shows both assimilation and contrast. Nevertheless, Luce et al. concluded that a one-back model works for all: "similar sequential effects are operating in magnitude estimation and absolute identification and ... in both cases they are confined entirely to the events of the previous trial."

Lockhead and I showed in boring mathematical detail just why a two-stage model – a model that takes account of at least two previous trials – is necessary to explain these effects[10]. But Luce's error persisted in the literature for many years – as far as I know (I've not kept up with this line of work) it persists still. As the rediscovery of Mendel shows, even in science you may have to say the same thing several times before the community pays attention. The need to repeat has probably grown in recent years, possibly because the supply of new ideas has failed to keep pace with the increasing number of scientists.

Figure 6.3 Sequential effects: From Staddon et al., 1980.

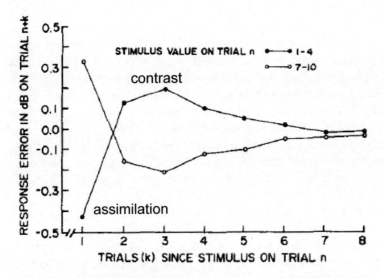

Figure 1. Response error on trial $n + k (k = 1, 2, ..., 8)$ following the presentation of Stimuli 7–10 (or 1–4) on trial n.

In addition to the AI course at MIT, I took a course on biological and neurological control systems with vision researcher Larry Stark, an MD by training, who kept a caged live owl in the main workspace of the

---

[10] Staddon, J. E. R., King, M. & Lockhead, G. R. (1980). On sequential effects in absolute judgment experiments. *Journal of Experimental Psychology: Human Perception and Performance, 6(2),* 290-301.

laboratory. (Now that *is* an eye! But these days the bird would be prohibited by animal-care rules; humans and animals are not supposed to mix.) MIT was an exciting place at that time. Self-confessed 'Ex-Prodigy' Norbert Wiener had coined the term and identified the new field of *cybernetics*. Hopes were high that real 'artificial intelligence' was just around the corner. A computer will be world champion chess master "within 10 years!" predicted Herb Simon at Carnegie Mellon. (In fact it took 30 to 55 years, depending on your criterion.) Jerry Lettvin and his collaborators had just published "What the frog's eye tells the frog's brain" a groundbreaking account of the 'front end' of neural processing in the visual system. Eccentric psychiatrist Warren McCulloch and reclusive mathematician Walter Pitts were writing papers with titles like "A logical calculus of the ideas immanent in nervous activity", which showed that a simple nerve-cell model could emulate a universal Turing machine.

The idea of feedback control was pervasive and Larry Stark was one of the experts. His specialty was the neurophysiology of vision. He had devised various ways of controlling light input to the eye, independently of pupil size, so as to excite pupillary movement and allow identification of system properties. I did two tiny projects in Stark's lab as part of the course. I worked in a corner, recording the response of crayfish optic nerve to various intensities of light (how dull is that!). I found that the rate of nerve firing was related by a power function (!) to the intensity of illuminating light. I also did a psychophysical experiment in Mem Hall which showed that visual acuity for black on white letters is better than acuity for white on black. The experiment was hardly necessary since the conclusion should have been obvious. These projects were my only Cambridge contact with Smitty Stevens' work. They were not worth any attempt to publish.

My second project for Stark was modest experiment on timing, which was to become my main interest. It involved a handful of human subjects and a small – i.e. about as big as two large refrigerators! – computer in Stark's lab. (Integrated circuits had only just been invented and had not come into general use.) People and animals can learn to wait in between individual responses, such as pressing a lever or tapping a computer key, if some kind of reward depends on the wait. I wanted to see (a) if human subjects (approximated by MIT undergraduates) could learn to do this; and (b) how they would treat the other keys on the keyboard. The subjects were told that they could earn points by pressing on the space bar. There were no other instructions. The *reinforcement schedule* was something called spaced responding, or differential reinforcement of low rate (DRL). Only presses of the space bar separated by more than 10 seconds earned a point.

In the short time each student worked at this task, almost all developed a pattern of taps on other keys, in between presses on the space bar, that took long enough to wait out the 10-s requirement. They thought they had figured out the rule. As I recall, only one or two realized that it was just time that was critical. This little experiment touched on what was to become a hot issue in the animal-learning literature: how do animals tell time? Do they just develop a stereotyped pattern of behavior sufficient to fill the time? Do they have some kind of ticking 'internal clock'? Or do they adapt to temporal regularities in some other way?

During my last year, I took the only actual biology class of my career, with Donald Griffin. It was on animal behavior. Griffin is of course famous for his pioneering work on the echo-location capabilities of bats. In his later years he became interested in animal consciousness, a line less scientifically fruitful. To complete my attempt at divorce from real psychology I took, also at Harvard, a demanding, excellent, course in electrical engineering with a splendid text by H. H. Skilling. I passed the course but with no great distinction.

There were research talks – colloquia – almost every week. Graduate students were deputed as hosts for the various luminaries. The one I remember most vividly, because I was the host, was famed mathematical linguist Noam Chomsky, from MIT. Chomsky was mainly a linguist in those days. He was soon to become known for a destructive critical review of Skinner's darling: *Verbal Behavior*, published a year or so earlier. In fact the review was a largely misdirected, but nevertheless effective, critique not just of *Verbal Behavior* but of behaviorism *entire*. Skinner, fatally, declined to respond, so Chomsky's charges stuck. Now Chomsky is famous as an activist, anarcho-syndicalist (something like the Burgess Hill system, I think) and all-around critic of the West, especially the United States. He has become one of the world's most famous public intellectuals.

Chomsky is a brilliant rhetorician. I was particularly struck recently when he placed the great 18th century free-market economist Adam Smith in the egalitarian, even socialist camp (although socialism was not yet born). Who knew! But there are quotes that seem to make Chomsky's point. For example, from *The Wealth of Nations*:

The whole of the advantages and disadvantages of the different employments of labour and stock, must, in the same neighbourhood, be either perfectly equal, or continually tending to equality. If, in the same neighbourhood, there was any employment evidently either more or less advantageous than the rest, so many people would crowd into it in the one case, and so many would desert it in the other, that its advantages would soon return to the level of other employments. This, at least, would be the case in a society where things were left to follow their natural course, where there was perfect liberty, and where every man was perfectly free both to choose what occupation he thought proper, and to change it as often as he thought proper.

Smith seems to assume that all men are potentially equally competent at all professions – that the human mind is "a piece of unformed wax, moulded entirely by conditions acting after birth" – which was the general belief in the 18th century, according to Cyril Burt. Which allows Smith to say that perfect liberty will lead to perfect equality, a claim precisely the opposite of the usual conclusion. Given the differences among men, perfect competition will inevitably leave some higher than others. *Liberté* and *Egalité* don't get along well together. I can't believe that a modern Adam Smith would disagree.

Smitty Stevens' power law held sway over the eastern end of Memorial Hall, but R. J. Herrnstein's *matching law* dominated at the other end. To understand this law, I need to explain how Skinner's conditioning experiments were done. Operant conditioning is just the delivery of a reward for a response, for some measurable thing the animal has to do. In Mem Hall, the popular animal was a pigeon, the response pecking a lighted translucent disk (called a key) and the reward a few seconds access to food. The pigeons were kept at about 80% of their free-feeding weights, to ensure that they were hungry[11]. Usually the reward – now termed a *reinforcement* – follows immediately on a key peck.

Skinner discovered early on that an animal will keep trying, even if you don't pay off every response: read his wonderful paper *A case history in scientific method* for an account of his discovery[12]. This led to his idea of *schedules of reinforcement*, different ways to arrange for the peck-reward *contingency* (Skinner's word; a better would have been *dependency*). For example, instead of requiring one peck for each bit of food you might require ten, so only the tenth peck in a run is paid off. This is called a Fixed-Ratio (FR) schedule. Or, if only the first peck more than 30 s after the previous food delivery gets rewarded; this is a Fixed-Interval 30-s (FI 30) schedule. The time doesn't need to be fixed; it can be variable, even random, so that 'set-up' is determined by a biased coin tossed every second. If the coin comes up heads, it stays so until the next

---

[11] "But it's cruel to keep them so hungry", you cry. Well no: a wild-caught pigeon is at about 80% of the weight it will reach with free feeding under lab conditions.

[12] Skinner, B. F. A case history in scientific method. *American Psychologist*, Vol 11(5), May 1956, 221-233.

peck, which is paid off, after which coin-tossing resumes. If the coin comes up heads on average once every 30 s, this is called a Variable-Interval 30-s (VI 30) schedule.

Animals adapt easily to these different arrangements. When food deliveries are at least 30-s apart (on a FI 30 schedule), the bird doesn't start pecking immediately after food, but waits perhaps 15 or 20 seconds before beginning. On FI 60, he waits 30-40 s and so on. On a VI schedule, food can come at any time. Pigeons behave accordingly, responding at a steady moderate rate.

The animal's behavior was made visible in real time, by another of Skinner's inventions: the *cumulative recorder*. Time was on the horizontal axis, and each peck pushed the record up by one unit on the vertical axis. Changes in response rate could be seen at once as changes in the slope of the record.

Skinner argued that response rate is a critical measure of the response probability, of *response strength*. The steady responding on VI schedules seemed to provide a handy measure. This emphasis on an average rate, as opposed to real-time changes – on counter totals rather than cumulative records – has turned out to be misleading in some respects.

The matching law arose in an experiment in which the pigeon faced not one but two response keys, each associated with its own VI schedule (termed a *concurrent* VI VI). An essential feature of VI schedules is that the probability of payoff depends on *not* responding. The longer the pigeon waits to peck, the more likely the coin has come up heads and his peck gets food. At the very least, this surely acts to prevent the pigeon in a two-choice situation staying too long on one key. Even if key A, say, pays off at twice the rate of key B, the more time he spends pecking A, the more likely that his next peck on B pays off.

The matching law is just this: The ratio of responses (totaled over a few experimental sessions, typically five hours or so) to the two keys matches the ratio of reinforcements obtained: if the left/right reinforcers are in the ratio 3 to 1, so also will be the responses.

Herrnstein and his students thought the law a very general principle that could help us understand how rewards work, not just in the animal laboratory but also in human society. But I was suspicious, simply because the procedure is so complicated. Smitty Stevens' power law had been criticized because his procedure, just asking people to give a number, seems so uncontrolled, even arbitrary. The results, at least across a small group of subjects, are orderly and justify the method. But the more complex the procedure, the more likely that it has been unconsciously tailored to produce not necessarily a specific result, but a result that fits the researchers preconceptions. After all, well before Herrnstein's experiments in the 1950s and 60s, very simple studies had shown that when given a choice, animals will usually go exclusively for the best – biggest, highest-probability – option. Not matching, but straightforward maximizing. This led in later years to more or less successful theoretical attempts to reconcile the two: matching as the outcome of maximizing. But why then pay so much attention to a procedure that obscures what is really going on? I return to this issue later.

Back to choice between two VI schedules (termed *concurrent* VI). What do the pigeons actually do? In fact they don't usually match. Instead they either just alternate between the two keys or *undermatch*: if the reinforcement ratio is 3 to 1, the response ratio may fall short, just 2 to 1, for example. Not perfect matching, then.

Herrnstein saw undermatching as a problem and thought he saw a solution to it: the 2-choice concurrent VI experiment actually involves (he claimed) not two choices but three. The third choice is *switching* from one key to another. To test this idea, Herrnstein modified the procedure. He added a further complication, what is called a *changeover delay* (CoD): neither response could be rewarded for a second or two after each switch from one response to the other. With a CoD added, Herrnstein got exact matching[13].

Herrnstein had no independent evidence that switching is a 'third response'. After all, the animal would switch even if responses were randomly allocated to each key. And the CoD isn't a *test* of anything. It is a procedural modification that has a number of effects. It certainly discourages switching, since the first switch response is never paid off. It enhances the "if you wait things get better" property of VI schedules, because if you switch less, the payoff probability when you do switch (albeit after a 1.5 s delay) increases. The real justification for the CoD was that it worked: it produced orderly results that fit in with what most people then expected. Unfortunately, later work showed that if the CoD is too long, the result is not matching but *overmatching*: if the reinforcement ratio is 3 to 1, the choice ratio might 4 or 5 to one[14]. Nothing special about matching, then?

Here is the problem: in all operant-conditioning experiments, the pigeon is just one half of a feedback system the other half of which is the reinforcement schedule. The steady-state of the *system*, be it alternation, matching or undermatching, may reveal little about what is going on in the *pigeon* which, to me at least, was the point of the whole exercise.

---

[13] Herrnstein, R. J. (1961) Relative and absolute strength of response as a function of frequency of reinforcement. *Journal of the Experimental Analysis of Behavior*, 4, 267-272.

[14] Baum, W. (1982) Choice, changeover and travel. *Journal of the Experimental Analysis of Behavior*, 38, 35-49.

The backdrop to all this was Skinner's emphasis on the sovereignty of response rate, which was never really challenged. Not challenged, even though the fact that rate itself could be controlled by reinforcement – by, for example, the 'differential reinforcement of low rate' (DRL) schedule – was already well known. Plenty of experimental results showed that response rate is not in any sense a pure measure of 'response strength'. Nor is the response-strength concept without its own problems, as I'll explain in a moment. So, given the arbitrary cognitive assumption that switching is a response just like pecking a key, the questionable status of response rate as a measure of response strength, and the dodgy theoretical status of response strength itself – there really was reason to suspect that the matching law might not be as fundamental as many Mem Hall inhabitants seemed to assume. Perhaps it is nothing more than another ingenious demonstration of the power of a well-crafted reinforcement schedule to produce orderly results – and *order* was another Skinner mantra. He was fond of quoting Pavlov: "Control your conditions and you will see order".

Here's a simple example to show what's wrong with the idea that response rate on a variable-interval schedule is an adequate measure of the 'strength' of the response. Imagine two experimental conditions and two identical pigeons. In Condition A, pigeon A pecks a red key for food reward delivered on a VI 30-s schedule. In Condition B, pigeon B pecks a green key for food reward delivered on a VI 15-s schedule. Because both food rates are relatively high, after lengthy exposure to the procedure, the pigeons will be pecking at a high rate in both cases: response rates – hence 'strengths' – will be roughly the same. Now we change the procedure for both pigeons. Instead of a single schedule, two schedules alternate, for a minute or so each, across a one-hour experimental session. The added, second schedule is the same for both pigeons: VI 15 s, signaled by a yellow key (alternating two signaled schedules in this way is called a *multiple* schedule). Thus, pigeon A is on a *mult* VI 30 VI 15 (red and yellow stimuli) and pigeon B on a *mult* VI 15 VI 15 (green and yellow stimuli). Unsurprisingly, B's response rate in green will not change. All that has changed for him is the key color – from red all the time to red and yellow alternating, both with the same payoff. But A's response rate in green, the VI 30 stimulus, will be much depressed, and response rate in yellow for A will be considerably higher than B's yellow response rate, even though the VI 15-s schedule is the same in both.

The effect on responding in the yellow stimulus by pigeon A, an increase in response rate when a given schedule is alternated with a leaner one, is called positive *behavioral contrast* and the rate decrease in the leaner schedule is negative contrast. Behavioral contrast was first described in an important experiment by George Reynolds that came out while I was at Harvard. But the main conclusion, which should have been obvious, is that response rate alone is inadequate as a description of the 'strength' of an operant response.

J. A. "Tony" Nevin many years later suggested that a property of operant responding in addition to its rate is its 'resistance to change', or 'momentum', although neither term quite captures what is happening[15]. Subsequently, I suggested the term *competitiveness* to describe effects like contrast because all the activities in an animal's repertoire compete for the available time. A reinforced activity is more competitive if it is reinforced more frequently. The next section (a bit technical, sorry!) shows how this works.

Matching and contrast

The theory I am about to sketch has two assumptions. First, that the rate of an operant response depends on two things: the reinforcement rate for it, and the strength of competing activities. Second, that there is always competition, even if only a single response is reinforced – on a single-key variable-interval schedule for example.

Figure 6.4

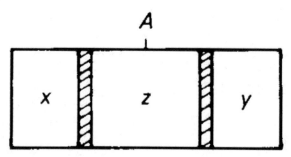

Figure 6.4 embodies these assumptions in a simple physical analogy that represents choice between two variable-interval schedules. It depicts an airtight cylinder of unit total volume, with two 'free pistons' that move until the pressure is the same in each of three compartments, with volumes x, y and z. Let volume represent number of responses x and y made to each choice; z is the notional competing behavior. Now suppose that the number of gas molecules corresponds to the reinforcement rates, R_x and R_y – the more molecules the higher the pressure – which is the ratio of the

---

[15] Because the supposedly resistant-to-change higher-reinforcement-rate component actually changes almost as much as the low-resistance to change lower-reinforcement-rate component in the contrast experiment: its rate increases as rate in the other component decreases.

number of molecules to the volume. Boyle's law implies that the pistons should move until pressure is equalized. In which case $R_x/x = R_y/y$, which is matching.

Now suppose we eliminate response y, to simulate the single-response case. The relation now is $R_x/x = R_z/z$. Since volume z is, by convention just $1-x$, rearranging yields $x = R_x/(R_x+R_z)$ which, since R_z is assumed to be constant, is just Herrnstein's well-accepted relation for the single-response case[16]. Notice that if both R_x and R_y are large in relation to R_z, response rates x and y will be similar even if R_x is greater than R_y, as in examples A and B above.

Behavioral contrast is only slightly more complicated. The distinctive feature of a multiple (two or more successive components) schedule is that the responses in each component can only take up part of the total time – half each for an equal-component 2-component schedule. This is represented in the diagram by a 'detent' in the middle of the cylinder a (*A* in the figure) which limits the travel of the pistons so that neither x nor y can exceed 0.5. Now, consider what happens when we go from $R_x = R_y$ (condition A in the contrast example above) to $R_x = 2R_y$ (condition B). If R_z is negligible, then in the first case, when $R_x = R_y, y = R_y(R_x+R_y) = R_y/2R_y = 0.5$. In the second case, after R_x is doubled, $R_y = R_y/3R_y = .33$ a reduction in y. On the other side, $x = 2R_y/(2R_y) = 1$, an increase in x: positive contrast.

The analysis is a bit more complicated if R_z is significantly greater than zero – contrast is in fact maximal when $R_x = R_z$. If R_z is large enough that neither x nor y can 'saturate', take up the whole time in their side of the cylinder, then you get matching; otherwise you get undermatching. All this corresponds pretty well to the available data[17]. This simple model can be easily tested. For example, it predicts that if the duration of the rich component is reduced from – 0.5 to 0.2, say – response rate in the lean component should increase, which seems to be the case[18].

Conclusion: Nevin's 'momentum' and Herrnstein's matching result both follow the same law, so long as the time available for each response is taken into account. And the 'state' of the pigeon requires at least two numbers to specify: both response and reinforcement rates. Response rate alone is not sufficient.

---

[16] Herrnstein, R. J. (1970) On the law of effect. *Journal of the Experimental Analysis of Behavior*, *13*, 243-266.

[17] All explained in more detail in *Adaptive behavior and learning* 2nd edition, Cambridge University Press, 2016.

[18] Behavior competition, component duration and multiple-schedule contrast. Ettinger, R. H.; Staddon, J. E. *Behaviour Analysis Letters*, Vol 2(1), Jan 1982, 31-38.

What does all this say about the process operating in the individual pigeon? After many years of work by many people, trying to understand the process that underlies matching, the answer in brief is: not much. In 1983, John Hinson and I tested the idea that the response rule followed by the pigeon in concurrent VI VI choice experiments is almost irrelevant: matching happens, no matter what[19]. We simulated a pigeon that followed a weighted combination of two rules: choose the response made least recently (i.e. switch) and the Law of Effect (Just reinforced? Respond a bit more.). The schedule was simple concurrent VI VI (no CoD). The L of E, named by learning pioneer Edward Thorndike in 1898, is simply the basic principle of reinforcement: repeat a just-rewarded response[20]. 'Least recent' is a simplified version of the payoff contingency on a VI schedule: the longer you wait, the higher the payoff probability. The simulated pigeon followed a weighted mixture of two strategies: switching and (reward-induced) perseverating. We showed that no matter what the relative weighting of these two processes, the result was undermatching, much like actual pigeon data. We concluded that the fact of matching tells us very little about the underlying choice-by-choice process.

While our article was still in press, I made this point in a throwaway line at the end of a commentary on a 1984 article[21] by biologist John Maynard Smith on game theory: "Matching is not universal, depends critically upon the feedback relation between reward rate and response rate, and is predicted by a long list of learning rules." This caught Dick Herrnstein's eye and he wrote me a longish letter of rebuttal: "As far as I know, there is not one single clear case of dependence of matching on

---

[19] Hinson, J. M., & Staddon, J. E. R. (1983). Matching, maximizing and hill climbing. *Journal of the Experimental Analysis of Behavior*, 40, 321-31.

[20] Thorndike, not a behaviorist, interpreted the effect of reward not just in terms of the observed behavior but in terms of supposed 'associations' of which the behavior was merely an indicator. But I use 'Law of Effect' just to describe the immediate observable effect of reward in giving a boost to a behavior which it follows. The rule Hinson and I used in our simulation was very simple: The model responds every second, choosing the response (Left or Right) with the highest V value. V for each response is just $V = t + K/T$, where t is the time since the response last occurred and T is the time since it was last reinforced; K is the parameter we varied in each simulation. If, say, the left response is rewarded, then T for that choice will be very small just after reinforcement, so the right-hand term will be very large, which will usually lead to repetition of that response. In the absence of any reward, the second term will approach zero for both responses, so the simulation will alternate, choosing the response made least recently.

[21] Staddon, J. E. R. (1984).It's all a game: A commentary on J. Maynard Smith Game theory and the evolution of behavior. *Behavioral and Brain Sciences*, 7, 116-117.

feedback relations in the entire literature." Chastened, I responded, "[M]y main objection…is that it is hard to think of a choice rule sufficient to produce (a) nonexclusive choice on [concurrent] VI VI and (b) extinction to an unreinforced alternative, that does not also generate matching…matching seems to be a joint product of an animal so constructed that it roughly follows the course of reward, together with a feedback function that shows diminishing returns to effort. In other words, the principle tells us much more about the combined animal-schedule system than about the animal itself."

For an animal choosing between two VI schedules, matching is the outcome for quite a wide range of choice rules. So long as the bird is more likely to go left for a while after left reinforcement, and similarly for the right, matching, or something close to it, will inevitably result. It follows, therefore, that the concurrent VI VI situation is not very useful as a way to understand the choice *process*, what pigeons are really doing as they adapt to reinforcement. But I don't think Dick was ever convinced and work on matching continued to dominate operant choice research.

I had not worked all this out in 1962, but I had misgivings about the complexity of the situation. So, although Dick Herrnstein was my thesis adviser, I had no interest in working on matching. (And Dick, to his great credit, allowed me to follow my own interests unimpeded.) So my other work at Harvard followed a different line: on how pigeons adapt to temporal regularities.

Chapter 7: Time and Cognition

The Memorial Hall pigeon lab[1] was an exciting place in the early 1960s. The facilities were great and the lab itself ran like an industrial plant.

Pigeons, usually large male White Carneaux, bred for eating ('squab') in the Palmetto Pigeon Plant in Sumter South Carolina, were put in their Skinner boxes one after another, usually 4 to 6 a day for each experiment. They were in the box for an hour or two, after which counters and timers were read and, possibly, a cumulative record looked over – although CRs were rather passé by this time, just used to check that the apparatus was working as intended. The bird was given grain 'up to its 80% weight' after it finished its stint.

I was slightly puzzled that we did not just catch wild pigeons to use in experiments. The reason was I suppose disease – the need to quarantine them – and perhaps variability: their age would necessarily be unknown. But other labs have used, for example, discarded racing pigeons – with no problems. A little variety would have been nice.

Experiments were run five days a week (you did it yourself on weekends) by an odd but efficient couple: Wally Brown, a young African American, and Mrs. Antoinette C. Papp, a formidable lady of German extraction who had once run some kind of confectionary business. The two are acknowledged in dozens of papers from that era. The eager experimenter just needed to wire up the controlling relay rack, label the counters to be read, and show Mrs. Papp or Wally the "on" button. With this wonderful system, even a lowly graduate student could have three or four experiments running simultaneously.

The most active faculty member at that time was the late and very much lamented George Reynolds, then an instructor. I found out much later that despite his balding pate, George was just a year older than I. Every week, or so it seemed, everyone would get in his[2] mailbox a green-covered paper just published in *The Journal of the Experimental Analysis*

---

[1] See William M. Baum, The Harvard pigeon lab under Herrnstein. *Journal of the Experimental Analysis of Behavior* 2002, 77, 347–355, and other reminiscences in that issue.

[2] Or "her"; half my class was female. Usually "his" is generic.

of Behavior (JEAB: the Skinnerian operant-conditioning journal) by George, either on his own or in collaboration with Charlie Catania, then a postdoc. Also circulating was a fat mimeographed jointly authored manuscript entitled "A quantitative analysis of the behavior maintained by interval schedules of reinforcement" (finally published in 1968) that was a veritable mother-lode of information on the role of temporal factors in reinforcement schedules.

Reynolds left after a few years, as Harvard followed its tradition of not promoting junior faculty. Herrnstein was a partial exception: he returned to Harvard as an assistant professor, but only after three years working with behavior analysis pioneer (Colonel!) Joe Brady at the Walter Reed Army Research Institute in Washington. Reynolds went first to the University of Chicago and then to UC San Diego, where my classmate Ed Fantino joined him and an enclave of operant research began to grow. George drifted into administration, textbook writing and the creation of the successful popular magazine *Psychology Today*. He died prematurely in 1987: "He was a man of brilliance, outstanding intellectual breadth, finely-honed wit, and great warmth. That he accomplished in one decade of productive research more than most psychologists in a lifetime, provides little solace to those of us who will miss his extraordinary presence[3]."

There was one final set of examinations before I could get down to dissertation research. The Preliminary exam – 'prelims' – covered the whole range of experimental psychology. There were several exams, which I passed, I guess – I remember none of them. I do remember the 'advanced' French oral exam, which I took in lieu of lower-level French and German. Skinner was the examiner and was forgiving.

So, on to research, but first we all had to learn how to program the electromechanical devices – relays, timers and counters – used to count pecks, present stimuli (usually colors on the response keys), and deliver food reinforcement. I taught the little course myself a year or two later, but when I arrived the process was still pretty informal. I do remember Charlie Catania showing a few of us some horizontal racks, each with a bunch of relay equipment. We were then left to wire up flip flops, lock-ups and the various other simple circuits needed for basic schedules. Things were not going well, when George Reynolds appeared and told us that some of the equipment on the old racks was faulty. No wonder we couldn't get things to work! He gave us access to newer vertical racks and equipment and things then proceeded more smoothly. The snap-lead method for rapidly wiring up a relay circuit was apparently introduced to

---

[3] Obituary: 1988, University of California: In Memoriam.

psychology by Skinner student Norman Guttman, a future colleague at Duke.

I was very much influenced by what I had been learning at MIT about feedback systems. I began therefore with an extraordinarily naïve experiment to see if reinforcement – the periodic delivery of a small amount of response-dependent food to a hungry pigeon – acted like a driving 'input' to the pigeon. The schedule was a sort of variable-interval, but the intervals were not random. They varied cyclically from short to long to short, for four cycles in a two-hour session. That was the 'input'. The output was simply the response rate over successive 5-min intervals (the response-rate zeitgeist still had me in its grip, for a while at least). The results were striking enough to get the paper, my first, into that Holy Grail of the aspirant researcher: *Science* magazine. I concluded "Daily exposure of pigeons to four cycles of a reinforcement schedule in which the density of reinforcements varied cyclically as a function of time induced a periodicity in their responding matching that of the schedule, but out of phase with it."

Bill Morse, a wise old student of Skinner's, at the Harvard Medical School, and a regular attendee at the 'pigeon staff' meetings, had a critical comment: "Maybe they are just pausing for a while after every reinforcement, so they spend proportionately more time pecking during the long intervals of the cycle" – which would of course result in a higher response rate at times of low reinforcement rate, as I found. I was also puzzled at the result. I expected higher response rates at times of higher reinforcement rate, not the reverse. I looked at the cumulative records to see if there was any evidence for the post-reinforcement pause Morse had suggested, but could see none.

But of course Morse was basically correct. The series of intervals that made up my cycle varied rather little. There were just a few long intervals. The pigeons therefore adjusted their waiting time to the short intervals, hence spent proportionately less time waiting during the few long ones, although this was not obvious in the cumulative records. The result: more pecks per minute in the longer, lower-reinforcement-rate parts of the cycle.

With a better data-recording technique, Nancy Innis, my first graduate student, was able to show a few years later that pigeons are perfectly able to track cyclically varying intervals[4] so long as there really is a

---

[4] Innis, N. K., & Staddon, J. E. R. (1971). Temporal tracking on cyclic-interval reinforcement schedules. *Journal of the Experimental Analysis of Behavior, 16,* 411-423. A more recent paper on this topic is Ludvig, E. A. & Staddon, J. E. R. (2005). The effects of interval duration on temporal tracking and alternation learning. *Journal of the Experimental Analysis of Behavior, 83,* 243-262.

progression, and not just a few long intervals alternating with a bunch of short ones. Moreover the tracking is in phase, because what tracks is not the response rate, but the *waiting time*, the time after reinforcement before the bird starts to peck. Usually the wait time in an interval is proportional to the interval just preceding. It turns out that time is often, perhaps always, a more useful clue to what is going on than response rate.

Nancy was a good friend as well as a student and we worked together over many years after she graduated from Duke, eventually to join the faculty of the University of Western Ontario in Canada. She very kindly organized a festschrift at Duke for me in my 65th year. The dinner was a grand occasion, held in the Doris Duke building in the Duke Gardens. My son Nick, gifted musically as I had hoped to be but was not, played the piano, and many students and postdocs from as far away as Japan, attended. I was presented with nice-looking wooden box containing the manuscript of a book with papers given as part of the festschrift. The book was published[5] a few years later.

Nancy died suddenly and unexpectedly of a brain hemorrhage while visiting the 2000-year-old Buddhist temple of Yumbulakhang near Tse Dang in Tibet in August of 2004, at the young age of 63. The 12,000 ft. elevation of the monastery may have contributed. Nancy Innis was a true scientist, interested in finding the truth; not a careerist, interested in becoming famous. She died as any of us would wish, quickly and without pain – but much too soon. The ambitious biography of Berkeley psychologist Edward C. Tolman which she had worked on for many years remains unfinished.

I had to do a little teaching as part of my assistantship. Teaching was not my strong suit. The reason was not bad preparation but timidity. In England, we never had to give presentations to a class in school. Talks at university were restricted to reading an essay to a small tutorial group. So when my assistantship required me to teach relay programming to the next class of graduate students – a total of six! – I divided them up into three groups of two, which I taught separately.

I was really not ready for my first large class: 66 pharmacy students for introductory psychology at the University of Toronto, my first teaching job. Despite extensive preparation, I exhausted my material after twenty minutes into the one hour class. Things improved after few weeks.

Teaching relay programming led me a couple of years later to write a manual. Like most such exercises in recent decades, it was soon overtaken by the march of technology.

---

[5] Innis, Nancy K. (Ed.) (2008) *Reflections on adaptive behavior: Essays in honor of J. E. R. Staddon.* Cambridge, MA: MIT Press.

Twenty years later, the same obsolescence sidelined *The Countess Ada Lovelace Diet, Sex, Health, Workout and Database Book.* It was a silly-title little book on database programming in a program called dBASE II, to be published in Durham, NC, where I was teaching at Duke University. But the publisher was laggard. dBASE II, and the CP/M operating system in which it worked, faded before he got around to publishing the book. (Too late, I got a letter from Pitman in the U.K., who were interested and probably would have actually published *Ada* in a timely fashion.) dBASE II wasn't even "II". Apparently the author gave it the "II" label to suggest that it was revised and thus bug-free. It was neither. Relational data base is a nifty idea, though. I thought it might have some relevance to memory processes. In 2015, Countess Ada, a nineteenth century mathematical-logic pioneer and daughter of the anti-nerd Lord Byron, is right back in fashion, with Walter Isaacson's *The Innovators* and the feminist emphasis on women in IT.

I was a teaching assistant in Skinner's large Nat. Sci. 114 introductory

psychology course. No lecturing required, just answering students' questions and grading exams. The text was his reference-free *Science and Human Behavior.* Skinner obviously thought that most previous work in psychology – indeed on human society – had been rendered obsolete by radical behaviorism. He felt little need to acknowledge any antecedents.

He also had the students read three novels: *1984, Brave New World* and his own *Walden II.* I was familiar with the Orwell and Huxley books, but had to read *Walden II* which is a utopian account of a society run according to Skinnerian principles. Skinner's big idea was that society should pay little attention to established custom. We should constantly experiment; innovations should be judged by their success. Sounds good; sounds a bit like what most people do much of the time. But the big question, of course, is how do you measure *success*? Societies have many customs whose purpose is obscure, ranging from the seven-day week to proscriptions, religious and otherwise, about eating, clothing, prayer and the relations between the sexes. Many of these practices have endured for centuries: does that give them some credibility? Even safety standards, which seem to be objective, must strike some kind of balance between prudence and healthy tolerance for risk. How should that balance be struck?

Perhaps because I grew up in England, possibly because I did not yet share a certain naiveté about human nature and political systems that seemed then (not so much now) to be part of the American psyche – possibly because I had read more philosophy and economics than is usual for a psychology student – perhaps for all these reasons, I was deeply unimpressed by Walden II. I was not alone. The undergraduate students in the course, many very bright, answered some of the essay exam questions with grumpy objections to Skinner's simplistic utopian vision. We teaching assistants were sufficiently amused to stick a few them up on the walls of our office.

I had not read Thoreau's *Walden* so did not know why Skinner chose it for his title. I found out subsequently that Thoreau offered a sort of Rousseauean vision of a life designed from scratch to maximize… something – Simplicity? Isolation? Indifference to petty humanity? Thoreau was a wonderful observer of nature; that seems to be his main positive attribute. He also had a gift for the memorable phrase: "The mass of men lead lives of quiet desperation" – but that phrase, like many others of his, is deeply pessimistic, unverifiable and probably not true. He was a staunch abolitionist, not from compassion for blacks but because of a passionate belief in self-ownership. Like Rousseau, but unlike Edmund Burke, the great political theorist, he placed no value on tradition, on the Darwinian testing of established values by their persistence through many generations. He trusted only himself. A narcissistic isolate, he was also dismissive of anything outside Concord, including European civilization: "nothing new does ever happen in foreign parts" he wrote, not long after the French revolution. Skinner's choice of *Walden* as a title is an interesting commentary on the man himself.

I was at that time completely focused on science. I had no interest in the social and political issues raised by *Walden* and *Science and Human Behavior*. I didn't read Skinner's social-policy best-seller *Beyond Freedom and Dignity* when it came out in 1971. I only became interested in these topics much later on.

Why can't they wait?

Now, I was intrigued by the behavior of rats and pigeons on spaced-responding (Differential Reinforcement of Low rate (DRL)) schedules. Pigeons can learn to space their pecks 10 s apart if that is a condition for reward, but they clearly find it difficult. Yet they have no difficulty on fixed-interval schedules, adapting their post-food wait time perfectly to times between food deliveries as long as an hour or more. Rumor had it that pigeons could not learn to space their responses more than 20-s or so apart. No one had shown spaced responding at longer times, like 30 or 40 seconds. I resolved to try it. The result was a long experiment which showed that indeed, most pigeons can just manage a 30-s DRL schedule, but only after many weeks of training[6]. Rats can apparently do a bit better, but even they cannot sustain more than DRL 40. At 40 s or so, rats and pigeons cease to time at all, their reinforcement rate drops drastically, and they treat the schedule like a VI.

An interesting side-note is the name itself: 'differential reinforcement of low *rate*', given that the schedule specifies a *time*, not a rate. The choice of DRL as the preferred name shows how committed the field was to response *rate* as fundamental.

So why the big difference between DRL and fixed-interval (FI) schedules? Why do rats and pigeons have no difficulty timing on FI but are severely limited on DRL? Two things have turned out to be important, one of them congenial to radical behaviorism, the other not. First, the Law of Effect (LoE), the almost-automatic burst of further responses after a response that has been closely followed by a reward. This is a perfectly Skinnerian process although the details of how it works are yet to be fully worked out. Second, *memory*, the fact that a key peck is a great deal less memorable than 3-s access to food. Memory played no part in the conceptual schema then prevailing at the West end of Memorial Hall.

The LoE process makes DRL responding difficult because it causes the pigeon to repeat a reinforced response rather than stop and wait. LoE must be inhibited if the bird is to adapt. He can adapt in two ways: Simply by slowing down, but still responding more or less randomly, as on a VI schedule – which is inefficient because many *interresponse times* will be too short, and go unrewarded. Much better is to *time*, to space responses so that almost all are effective – which is what pigeons do if the required delay is not too long. Memory is involved in timing, because the pigeon must use the last response as a *time marker*, the thing to time from. He has to remember the most recent response. If memory for the response is overshadowed by the memory for the last food delivery – a much more salient event – which precedes it, effective DRL responding is impossible. Hence the breakdown of timing at DRL values longer than 30 s or so. I was to follow up both these leads in later years.

The DRL experiment raised the issue of *metastability*. A pigeon's stable behavior when first exposed to a given spacing requirement, say 20 or 30 s, did not reflect its behavior after more training with shorter times. Successful spaced responding at longer intervals required some training with shorter ones. Some *steady states* could be recovered an indefinite number of times; others could not. Obvious really, but at that time a steady state was a steady state, more or less.

---

[6] Staddon, J. E. R. (1965). Some properties of spaced responding in pigeons. *Journal of the Experimental Analysis of Behavior*, 8, 19-27.

The year before I arrived at Harvard, George Miller and Jerome Bruner fired the first shot in the counter-revolution against behaviorism. Miller is perhaps most famous for his paper 'The magical Number seven, plus or minus two', which summarizes experimental results showing the limits of human working memory. Jerome Bruner was a developmental psychologist best known for his book *A Study of Thinking* (with Goodnow and Austin) and for showing the effects of motivation on perception in children.

The Miller-Bruner incursion took the form of the Center for Cognitive Studies, located just a couple of blocks from Memorial Hall. I used to attend their little gatherings and learned much. But I was skeptical of their willingness to take terms derived from introspection and use them as part of a supposedly scientific explanation. Noam Chomsky, Skinner's nemesis and a strong influence, pointed out that people must be pre-adapted in some way to learn language. This is surely true. He had proposed a generative grammar embodied in a mysterious "Language Acquisition Device" as a key ingredient of the 'language instinct'. Several people tried to make the idea more concrete by suggesting computer-like algorithms, some of which could be tested experimentally. In the end all failed their tests, and the idea faded as psychology though it remains of some interest to linguists. But humans *are* different from non-verbal animals. They must possess something like Chomsky's language acquisition device even though his proposal for a system separate from other, more primitive learning mechanisms, doesn't work.

Cognition and the pigeon

I might have taken a hot idea from the Center for my thesis research, but in the end I chose a cognitive problem rather unimaginatively inspired by what I had learned as an undergraduate. *Knowledge of results* is an old-fashioned name for informative feedback. Sequential judgment, as I studied it later, at Duke, with Greg Lockhead, involves feedback in this sense. The subject is given a series of tasks – judging the loudness of a sound, for example. After he completes each one, he's told the correct answer, so he can see whether he has over- or under-estimated. Change – improvement, hopefully – in subsequent performance is then measured. As far as I knew, nothing quite like this had been tried with animals. Reward is 'knowledge of results', of course, though of a simple binary sort: reward=correct, non-reward=error – but the direction of error is unspecified.

For the task I again chose timing: DRL with *limited hold* (sorry about the jargon, but it's hard to avoid completely...) Limited hold just means that interresponse time (IRT) must be greater than, say 10 s, but also less than, say, 11 s: DRL 10 LH 1, which is an example of a general class of limited-hold schedules: DRL T LH $0.1T$: required minimum duration T with a window of $T/10$.

To test the effectiveness of feedback the subject must be given a succession of tasks. The DRL LH procedure provided a simple way to do that. I required the pigeon to track a changing value of T, to see how the presence or absence of feedback affected his ability to track T. The feedback for the pigeons was just a brief change in the key color right after each peck. There were three feedback signals: too short: a flash of green; correct: food reward; too long: a flash of red. Presumably a smart cognitive pigeon will soon learn to wait longer after a green flash and shorter after a red one. What's more, the effect of the feedback stimuli can be easily tested simply by reversing them. Instead of too-short IRT→green the bird sees too-short IRT→red, for too-long IRT→red he sees too-long IRT→green. If the bird has learned to increase his IRT after green and to shorten it after red, then when these error signals are reversed he should show unstable behavior: responding faster and faster or slower and slower. (I was able to show a feedback-reversal effect of this sort much later with a different timing procedure).

The cyclic DRL schedule varied from $T = 8$ to $T = 30$ s, in steps of 5 s. Each value was in force for 5 min. Two cycles made up a single experimental session.

The experiment proceeded in two steps. First I needed just to see if the feedback signals helped the birds master a very difficult task: comparing feedback and no-feedback conditions successively in the same animal. Second to see if reversing the feedback caused instability, as expected.

I used a new method, just invented by Donald Blough at Brown University, to be sure that they were tracking by varying interresponse time, rather than just slowing down or speeding up a random rate. The method used an oscilloscope, a simple electronic circuit and a Polaroid camera to record a whole experimental session (those days with Ivan and army-surplus electronics were beginning to pay off!). It worked like this: Each key peck did three things to the oscilloscope trace: made a brief flash (a dot on the film), reset the trace to zero on the y-axis, and moved it over a small amount on the x-axis. The trace moved upwards along the y-axis all the time at a controlled rate. The camera lens stayed open for the whole session. The result was a picture that showed every interresponse time across the whole experimental session. Some years later I used this basic idea to make a device called the zero-crossing analyzer, to display bird song (see Chapter 11).

With feedback, all three pigeons learned to track beautifully, varying interresponse time cyclically in phase with the changing DRL requirement (Figure 7.1). As expected, performance was a little worse at the longer T values.

What was the effect of the feedback? Not very cognitive, it turned out. One bird did fine even with no feedback at all. All the birds did a little worse when feedback was reversed, but none showed the kind of instability I had expected. The birds were not using the feedback response-by-response to speed up or slow down. But they were using it somehow, since their performance deteriorated without it or when it was reversed. I did another experiment a few years later to try and find out just how[7]. The results were conclusive only in showing that the birds do not use these feedback stimuli in the way human subjects might, as response-by-response feedback. These experiments much diminished any enthusiasm I might have had for cognitive explanations of animal timing behavior.

Figure 7.1 Three pigeons tracking two cycles of the spaced-responding schedule with feedback.

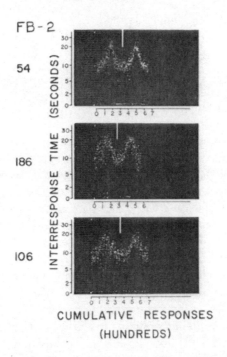

Donald Blough, the inventor of the oscilloscope IRT-recording method, was a student of Skinner's. He was a pioneer in several

---

[7] Staddon, J. E. R. (1969). The effect of informative feedback on temporal tracking in the pigeon. *Journal of the Experimental Analysis of Behavior, 12*, 27-38. (Reprinted in P. B. Dews (Ed.), *Festschrift for B. F. Skinner* (pp. 256-267). New York: Appleton-Century-Crofts, 1970) http://hdl.handle.net/10161/5994

directions. Perhaps his most creative contribution was a clever application to pigeons of a technique invented by Memorial Hall denizen Georg von Békésy to measure human auditory threshold. Think about it: a threshold is not as easy or obvious a thing to measure as, say, blood pressure or pH – because it is *private*. Only *you* know when you can't see a light or hear a sound. *You* can of course report what you see, but how do we ask a pigeon?

Blough's solution was to first train the bird to peck left (say) if a center key is lit and right if it not. After this task is well-learned, the brightness of the center key is slowly reduced. The bird pecks, LLLL and then switches to RRRR, as the luminance falls below his threshold. Luminance curves traced out in this way match human data, showing clearly the sharp transition from cone to rod vision.

The same technique could be used to see if a pigeon is subject to human visual illusions, like, the rabbit-duck illusion. Dick Herrnstein showed in a paper published the year I graduated that pigeons can readily be trained to recognize objects that fall into different categories – a Kodachrome slide of a scene with, or without a person in it, for example. Suppose now we train a pigeon with many target examples of rabbits (peck  left) and ducks (peck right). Once he's learned this task (shouldn't be too difficult for a smart pigeon), present the illusion figure. How will the bird behave: Will he respond at random? Will he just fixate on one side or the other. Or will he peck right for a while and then left? – which suggests that he is subject to human-like perceptual 'flipping'. For some reason, experiments like this don't seem to have been done, although some work has been done with a different method on the Zōllner illusion.

Dick Herrnstein died too young in 1994, of lung cancer, although he had not smoked for many decades (you mean you can get lung cancer *without* smoking!). He was a fascinating and important man. Very bright and willing to stick to his intellectual guns in the face of opposition. But also rather rigid (Rhoda's cat?) and possibly too captivated by Skinner's ideas, if not by Skinner. He never lost his devotion to response rate as a fundamental datum, for example, despite its arbitrariness – What is the right period over which to average? How do we know? And what about DRL schedules? – as well the other problems I've mentioned. Dick did publish a piece in 1977 critical of some aspects of Skinner's approach which elicited a grumpy response from Skinner and further rebuttal from him. Skinner rightly criticized Herrnstein's notion of self-reinforcement, but less forcefully than he should. After all, in self-reinforcement what is the contingency, the schedule? What is gained by attributing some apparently spontaneous behavior to an unidentifiable consequence rather than a potentially identifiable cause? Dick, in turn, rightly criticized Skinner's willingness to extrapolate from pigeon in Skinner box to man in society. Quoting Skinner: "Let him extrapolate who will. Whether or not extrapolation is justified cannot at the present time be decided" Dick goes on to say "His [Skinner's] subsequent writings leave no doubt that he felt he found ample justification for extrapolations to virtually every aspect of human affairs." Indeed.

Rumor had it that Skinner, even in his seventies, liked the ladies and would engage them with little restraint. One such was apparently Herrnstein's secretary and fiancée, a very attractive red-haired young woman, who was so favored both before and even after her engagement to Dick. I have no idea if this rumor is true, but there is no doubt that Herrnstein's article, and Skinner's rather haughty response – "When one has published nine books setting forth a scientific position, it is disconcerting to find it misunderstood. To be misunderstood by a former student and present colleague is especially puzzling[8]." – provides circumstantial evidence.

By the time I met him, Herrnstein was divorced from his first wife, Barbara. I didn't meet her when I was at Harvard, but much later Barbara Herrnstein Smith (if your last name is Smith, it is as well to add another) came to Duke as the Braxton Craven Professor of Comparative Literature and English. Barbara's interests are wide. She was interested in "theory", but when I asked her "theory of what?" I got no answer. In English departments, apparently, "theory" is *sui generis*, no need to ask. An interesting lady, she acknowledged no intellectual limits. Her literature students studied books like Gerry Edelman's *Neural Darwinism* and other semi-popular science, along with more conventional works, despite the students' uncertain backgrounds in biology or even science. She organized conferences to discuss papers like "Concepts: Constructing Quaternions" and "Complementarity and Idealization" – quaternions being a complex-number idea in mathematics and complementarity a concept from quantum mechanics. All this was stirred in a philosophical soufflé, with a Gallic *essence* contributed by Jacques Derrida, Jacques Lacan and their followers.

Jacques Derrida visited Duke in that period. He arrived, to a packed Page auditorium, Duke's largest, beautifully dressed in a grey-green double-breasted suit and elegant tie. "I 'ave been asked to read from my

---

[8] B. F. Skinner Herrnstein and the Evolution of Behaviorism. *American Psychologist*, December 1977, p. 1006.

book [*Specters of Marx*] for one hour" he said. "I will read for two hours…" I stayed for as long as I could.

In the 1970s, Dick Herrnstein began to write about human individual differences. In this also he echoed Smitty Stevens who had long been interested in W. H. Sheldon's *somatotyping*, the supposed relationship between body type – ectomorph, endomorph and mesomorph (roughly: thin, fat and muscular) – and personality traits. All in the tradition begun by Charles Darwin's brilliant cousin, Francis Galton, beginning with his book *Hereditary Genius* (1869). In a 20th century tripartite version of the medieval theory of four bodily humors, Sheldon thought that careful measurement of body type could yield three numbers, representing the contributions of the three types. He claimed that much could be deduced from these numbers about an individual's character and aptitudes. But the expected correlations failed to emerge. Sheldon's hopes were dashed and his method discredited. Nevertheless, his ideas were for a while sufficiently fashionable that as recently as the mid-1960s, Ivy League schools from Harvard to Vassar required undergraduates to submit to nude photography according to Sheldon's requirements. These 'data' were to allow him to assemble his *Atlas of Men* (which also included women). TV chat-show host Dick Cavett wrote an amusing account of his own experience as a subject in this bizarre exercise[9].

Dick Herrnstein was convinced by the overwhelming statistical evidence that intelligence – IQ – is largely inherited. He also assumed that American society is largely meritocratic: so that brighter, more energetic citizens will tend to rise up the socioeconomic scale. If human mating is *assortative* – like marries like, bright marries bright – then meritocracy implies that American society will eventually be stratified by intelligence[10]: 'brights' at the top, 'dumbs' at the bottom. In the last year of his life, he published a book with political scientist Charles Murray, who became a great admirer. *The Bell Curve* (1994) discussed race, class and IQ from this point of view and predictably inflamed the politically correct.

Dick Herrnstein very sadly died just as *The Bell Curve* was published. His co-author Charles Murray was invited to speak about the book at Duke the next year. I was asked to introduce him, a duty I approached with some trepidation. Murray had been treated disgracefully at several other institutions: heckled unmercifully, abused and boycotted. I thought very carefully about an introduction that might get the students actually to think about the issues raised by the book rather than just attempt to

---

[9] http://opinionator.blogs.nytimes.com/2011/12/02/last-nude-column-for-now-at-least/?_r=0

[10] R. J. Herrnstein. *IQ in the meritocracy.* Boston, Little, Brown, 1973.

extirpate it. My introduction began with a quote from the great 19th century observer of America: Alexis de Tocqueville: "I know of no country in which there is so little independence of mind and real freedom of discussion as in America." The rest of the piece is at the end of this chapter. In any event, it worked. Murray was received respectfully and his talk was followed by thoughtful questions from the audience.

Ability and social class is hardly a new issue. Great soul Dr. Samuel Johnson, in the 18th century, thought separation by inherited rank an excellent thing and meritocracy (although he did not use the word) a great danger: his biographer James Boswell writes:

> I mentioned a certain authour who disgusted me by his forwardness, and by shewing no deference to noblemen into whose company he was admitted. Johnson: "Suppose a shoemaker should claim an equality with him, as he does with a Lord: how he would stare. 'Why, Sir, do you stare? (says the shoemaker,) I do great service to society. 'Tis true, I am paid for doing it; but so are you, Sir: and I am sorry to say it, better paid than I am, for doing something not so necessary. For mankind could do better without your books, than without my shoes.' Thus, Sir, there would be a perpetual struggle for precedence, were there no fixed invariable rules for the distinction of rank, which creates no jealousy, as is allowed to be accidental."

So, for Johnson, differences of rank are not so bad, as long as they come about by chance: are "allowed to be accidental", like differences of height, looks – or ancestry. Men are not equal, said Johnson: "So far is it from being true that men are naturally equal, that no two people can be half an hour together, but one shall acquire an evident superiority over the other." Meritocracy implies constant struggle. When rank is decided essentially by lot, society is more peaceful – and also (though Johnson did not point this out) ensures that people of real ability will be found at every level.

Britain was more hierarchical 70 years ago. Our postman next-door neighbor in Cedar Road in the late 1940s used to play Bach on his piano. How many postmen now do the same? Of course today we believe in progress, and competition as its engine. Dr. Johnson didn't like competition – "perpetual struggle for precedence" – and did not believe in the possibility of continuous social improvement. The point is that hierarchy is not without benefits, nor meritocracy without costs.

The nature-nurture issue that got Herrnstein and Murray into trouble with *The Bell Curve* is a perennial one. Correlations – the basis for the heritability measures on which Dick's meritocracy argument relies – do not identify causes. The process of development, when nature and nurture are interwoven, is far from being understood even in animal models, much less human beings. Metaphors abound. Jonathan Haidt, for example, in an influential book, compares nature to the first draft of a book and nurture to the various edits and rewrites through which it metamorphoses into the final version. I tend to favor the metaphor of nature as akin to an instrument: some people are like trombones, others pianos or violins. Nurture plays the tunes. Arpeggios are tough on a trombone, chords hard to do on a violin. So the instrument limits and modifies what can be played on it. So nurture filters and modulates the effects of the environment. Don't like that metaphor? Think of another!

Back in Memorial Hall, after surviving some critical comments on my draft dissertation from one of the younger members of my committee, it was time to defend it. Skinner was the chairman of the committee. The examination went well. The only question I can remember concerned Harvard pioneer William James. It was about his book *The Varieties of Religious Experience.* I remembered only one thing about James and religion: that whenever he inhaled nitrous oxide gas to assess its consciousness-expanding effects, he had a revelation about the nature of the universe. But he could not recall it when the effects of the gas wore off. He determined, therefore, to write his revelation down the next time while he was 'under the influence'. He did so, and found that he had written: *Higamus hogamus, woman is monogamous, hogamus higamus, man is polygamous.* More than a grain of truth in it, but hardly a cosmic revelation. Whether the attribution to James is correct or not, I don't know. Doubts have been raised. But I passed the exam.

There was cute tradition in Memorial Hall. We didn't have phones in our offices. If you got a call, you would be paged over the PA system. The signal that a candidate had passed his PhD exam would be the announcement "Would Dr. Staddon please come to the Chairman's office." Thus it was I knew I had survived the ordeal.

CHARLES MURRAY INTRODUCTION

Duke University, March 2, 1995

"I know of no country in which there is so little independence of mind and real freedom of discussion as in America." These are not my words; they are the words of Alexis de Tocqueville, in 1835, in *Democracy in America*, a book that has more successfully captured the American character than any other, before or since. Although de Tocqueville admired much in this country, he was disturbed by what he called the *tyranny of the majority*, which he described in the following way:

"The authority of a king is purely physical, and it controls the actions of the subject without subduing his private will; but the majority possesses a power which is physical and moral at the same time; it acts upon the will as well as upon the actions of men, and it represses not only all contest, but all controversy..."

de Tocqueville's words are more than a 150 years old, but it is hard to imagine a more accurate summary of what is now known as *political correctness*. PC is a new variant on a very old theme, and it has universally censured *The Bell Curve*. As moral philosopher Michael Novak has said "On at least three matters – IQ, heritability, and human nature – the rules we have lived under for some decades now are evasion, euphemism, and taboo."

The book Charles Murray has written with the late Richard Herrnstein has been vilified to a quite extraordinary degree. Even before Dr. Murray's talk here at Duke, the organizers have evidently felt it necessary to arouse suspicion in the minds of those of you who have been residing for the past year in Outer Mongolia and failed to notice the blizzard of negative publicity that has surrounded the book. I can remember no other talk at Duke where the audience has been promised not just discussion, but *rebuttal*.

The Bell Curve has been chiefly criticized not so much for what it says about intelligence, class and race as for having the bad taste to discuss these subjects in public at all. Many members of the staff of *The New Republic,* for example, argued against printing anything about the book, even though the editor offered all an opportunity to respond. When they did publish a piece by Murray – preceded by several critical commentaries – it was clear that a number of critics had managed to protect at least one reader from the book, namely themselves. Other commentators did read the book, but decided to attack instead a work of their own invention. Influential science writer, paleobiologist, Marxist and (I am a little embarrassed to add) honorand of Duke University, Professor Stephen Jay Gould, for example, terming the book "anachronistic social Darwinism," in his *New Yorker* review, also wrote: "Herrnstein and Murray's second claim, the lightning rod for most commentary, extends the argument for innate cognitive stratification to a claim that racial differences in IQ are mostly determined by genetic causes..." later adding "Herrnstein and Murray violate fairness by converting a complex case that can yield only agnosticism into a biased brief for permanent and heritable difference."

Gould's use of the term "agnostic" is curious, because what Herrnstein and Murray actually say is

"It seems highly likely to us that both genes and environment have something to do with racial differences. What might the mix be? We are resolutely agnostic on that issue; as far as we can determine, the evidence does not yet justify an estimate."

Gould's misleading comments are just one example from an extraordinary barrage of criticism, ranging from ad hominem sneers: "Murray, an intellectual snake charmer. . . .(Tom Morgenthau, *Newsweek*)", through guilt by association: "The text evokes the dreary and scary drumbeat of claims associated with conservative think tanks: reduction or elimination of welfare, ending or sharply curtailing affirmative action. . . . (S. J. Gould again, in *The New Yorker*)." and open abuse: *The Bell Curve* "is just a genteel way of calling somebody a nigger" (Bob Herbert, *NYT*, quoted approvingly by psychologist Leon Kamin in a *Scientific American* review) to faint praise "I found it a much better book than I had been led to expect (Steve Blinkhorn, *Nature*)." The book has garnered a couple of A's –"This brilliant book (Father Richard John Neuhaus)" and "magisterial (James Q. Wilson)"– most in a thoughtful set of reviews in *The National Review*.

Well, after you hear what Charles Murray has to tell us tonight, I hope that unlike some of the book's more rabid critics you will go out and actually read it. Try and understand the issues; pay most attention to facts and arguments. Pay as little attention as you can to personalities and paranoia: neither the establishment nor the anti-establishment has a monopoly on truth. The main thing to remember is that more we know about this or any other controversial topic, the less we will be threatened by it.

Dr. Charles Murray was trained at MIT in political science. He first came to the notice of a wider public with his book *Losing Ground* (1984), an account of unintended consequences in our welfare system. He is currently a Bradley Fellow at the American Enterprise Institute, one of those "conservative think tanks" so alarming to Professor Gould.

John Staddon

Chapter 8: Oh, Canada!

I passed the PhD exam in January of 1964. No more tuition to pay, but I needed a job for the next few months until I could start teaching in the Fall. I had no doubt that I would get some kind of teaching job. It was definitely a seller's market in those days – lots of jobs, not yet so many PhD graduates. James Holland, co-investigator with Skinner on something called the Committee on Programmed Instruction (COPI[1]), came to my rescue and offered a postdoctoral fellowship for a few months. COPI was studying how to program educational materials. One product, a few years earlier, had been *The Analysis of Behavior: A Program for Self-Instruction,* a programmed text on the Skinnerian approach to learning.

Interest in programmed instruction – teaching machines – was probably at its peak in 1964. It has since faded to invisibility, despite the development of personal devices that would permit much cleverer programs than was possible then. My own little project hints at why. I decided to see if learning from a book could be improved by means of a simple reinforcement technique. I chose an excellent history book: Bronowski and Mazlish's *The Western Intellectual Tradition* (Harper, 1960). Bruce Mazlish was a historian at MIT, Jacob Bronowski a Polish-English polymath, and later writer and presenter of a wonderful BBC TV series *The Ascent of Man* (1973). The idea was to see if random questions popping up as a student was reading a text would somehow reinforce learning.

The experiment was a failure. Students thus prompted did no better than when reading on their own. They did fine just flipping from place to place as their interests dictated. I daresay they came up with their own questions and were just annoyed by mine. Either the book or the students were too good, or the basic idea was wrong. Well, probably all are true to some extent. The pigeon-shaping model – a well-defined response that you know the birds can do (like turning around), an artificially motivated (i.e. hungry) subject, and no issue of comprehension or 'insight' – simply did not apply to bright kids reading interesting but thought-provoking material. This was an early hint that the secret to teaching was Darwinian: not so much *selection* – reinforcement of behavior that just occurred "for other reasons" – but *variation*, getting the required behavior to occur for the first time. Improving teaching is about trying to understand those "other reasons". By interrupting the normal reading process I had

---

[1] James G. Holland and B. F. Skinner (Editors). An analysis of the behavioral processes involved in self-instruction. Title VII Project Number #191. National Defense Education Act of 1958.Grant Number 71-31-0370-151.3FINAL REPORT. Principal Investigator: B. F. Skinner. Not dated, but probably 1964.

probably impeded the spontaneous curiosity of bright students. Answering your own question with your own efforts is a great deal more 'reinforcing' than anything extrinsic I could offer.

I had graduated early, so I was the first of my cohort to seek employment. A couple of jobs came up. The first was at Dalhousie University in Halifax, Nova Scotia. Nova Scotia! Where the heck is that? Well, it is in Canada and I was later to have wonderful collaboration with one of the faculty there. But at that moment it seemed remote, and Dick H thought I could do better. The next possibility was also Canada: the University of Toronto. Well, at least I knew where that was. So I applied and was invited to visit.

The Psychology Department was on several floors of a modern building on St. George Street in the middle of Toronto. I met some people during my visit but, unusually, was not asked to give a talk – which was just as well as my lecturing ability was pathetic then and for a few years afterwards. But it was a symptom of the market situation, which favored the candidate. Perhaps it was also fatigue? Toronto Psychology was obviously expanding. They had hired five new faculty the year I visited and were to hire two more the next year. I was one of the two.

Which made things simpler than they might have been. The terms of my U.S. student visa required me to leave the country for at least two years before I could return as an immigrant. I did have a wife who was a U.S. citizen, which would probably have allowed me to overcome that limitation – after much bureaucratic grief. More recently, a young U.S. colleague of mine, with a new Mexican wife, had to wait a year and a half before she could legally join him in Durham. (I advised him to just have her come as a visitor, but he was a cautious fellow and the legal implications of that would apparently have been dire – even though she was in effect guaranteed a green card by virtue of her marriage. I'm not sure I would have been as patient.) But, as I accepted the Toronto offer, the problem never came up.

I disposed of my crappy car, but not before, finally, figuring out why the carburetor would flood and stall it from time to time. Too late! We rented a van to take our stuff up to Toronto, towing my new little car behind. Included were four pigeons from an experiment I had been working on in Memorial Hall. I remember some discussion at the border about the paperwork for the birds. I wondered at this, since the pigeons flying overhead seemed able to immigrate quite freely. Once again, evidence that there really is no limit to the studied obtuseness of the bureaucrat.

We had bought a new car. I was impressed by the engineering innovations – front-wheel drive, rubber suspension, transverse engine, etc. – of the British Motor Corp. The Mini, then just out for a year or two was in fashion with the intelligentsia. Possibly a brief breakout of long-dormant Brit patriotism was also involved. But the Mini seemed a bit small for America, so I went slightly upscale and bought a new BRG-colored MG 1100 – same design but larger. It drove like a dream, but the steering was a bit sensitive for someone used to squishy U.S. autos – as was my wife, who a month or two later over-corrected, fishtailed, and put the car into a spin in the middle of a freeway. Luckily, the other drivers saw what was happening, gave the car a wide berth and no one was hurt.

The new MG was little improvement over the old Ford. Mechanically, the car was truly awful. The electric fuel pump was just bolted under the trunk and attracted all available water and road debris. I had to replace it twice in two years. After that, I got rid of the car. In Toronto I replaced it with an enormous secondhand, white, Chrysler Windsor (that's 'Windsor, the other side of the river from Detroit', not 'Windsor, where the Queen lives'). Power steering, push-button automatic (of course – it was a North American car, after all) with a square steering wheel. I was a bit apprehensive at first because of the size (the 1100 could just about fit in the cavernous trunk), but in fact it was easy to manage and incredibly comfortable. Even the lack of air conditioning was not a problem. In the days before pollution control, and with no a/c, the engine compartment left plenty of room around the V-8 and allowed maximum airflow: two huge floor vents provided ample ventilation in the cabin.

We arrived in Toronto on May 11. There was not a leaf on the trees. This was a bad sign. Still the city was modern and seemed to work well. We rented a two-bedroom apartment in Don Mills, a new suburb of Toronto. The area was pleasant, though distant from St. George Street. One negative: a road leading to the freeway on the way into town was a revenue source for the traffic police. They used to catch cars freewheeling down the hill above the speed limit. This seems to be a universal trick Years later, I saw the same thing in Germany, for cars coming off the autobahn – where they might have been traveling at 80 mph or more. Traffic came down a hill, slowing, but not fast enough. The cops waited in the middle and fined you on the spot. Nothing on your record, though. The fine was basically a tax.

I didn't have to start in Toronto until the Fall, so that summer we went to Britain to see my family. We attended my sister's wedding in South West Lancashire, and rented a car to tour a bit of France. DB met my parents for the first time, as they were on leave from what was about to become Zambia. They soon returned, my mother working for a while as secretary to one of the ministers in Kenneth Kaunda's government: Reuben Kamanga. "Charming man, but a complete crook!" was her summary. My mother was not only a very bright woman, but also (family phrase) 'pathologically honest'.

My cousin Ben was then working as a film editor. His job that summer was on the third James Bond movie in a franchise that continues still. The movie was *Goldfinger* and Ben showed us around the set at Pinewood Studios just outside London. We saw the mockup of Fort Knox, where the U.S. stores its gold, when it has any. (I'm not kidding about the gold. Libertarian Congressman Ron Paul in 2010 questioned the accuracy of the government's audits.) And we saw "rushes" in which the Korean killer Oddjob flings his lethal bowler hat – over and over again, in the clips – at Sean Connery. The fight scenes were incredibly slow – all speeded up for the final version, of course. Ben did one or two more big-time editing gigs, but then I think reverted to producing video ads.

Another Don Mills negative soon made an appearance: construction of a large apartment building on the other side of the street. I first noticed a problem one day when I heard a very loud whirring noise. I looked out into the street and could see nothing untoward. I went downstairs and on the ground floor I saw a car with its engine idling. I could barely hear it. Evidently the just-completed 8-storey building opposite had focused the sound so that it was much louder in our fourth-floor apartment than at ground level.

I managed to make a take-apart-able bookcase in the kitchen of our Don Mills apartment, but it wasn't very convenient either for woodworking or commuting. So we moved to an older apartment, with a little deck at the front, on Lowther Avenue, a few blocks from the Psychology Department. We rented from a man called Mark Stein. He insisted that he was just a representative, not the actual landlord, which meant that we could not hold him personally responsible for recurring defects in the old building. Just before we left Toronto, two years later, we saw Mr. Stein driving by with his family – in a new Rolls-Royce. Apparently he *was* the landlord and had made a bucket of money in Toronto's booming property market.

Still, Lowther Avenue was a convenient location and I could walk to the Department, which I did even in the depth of winter. I once or twice felt sure I would perish in the cold walking down Bloor Street, saved only by nearby shops in which I could seek refuge. Montreal has adapted to the Canadian climate by having much of its central shopping area underground. Toronto had failed to take that step.

In our second year in Toronto, our son Nick was born. DB had wanted to have children while we were still in Cambridge, possibly because Shaler Lane was full of tricycles, toddlers and baby-buggies. But it was to be Toronto. Neither of us wanted me in at the birth. Marital division of labor (no pun…) had not yet broken down. So Nick appeared in a Toronto hospital attended only by women. He was a happy, healthy babe and was doted upon appropriately.

In Lowther Avenue we purchased the dreaded AirCrib, an invention of Skinner's that has been transmogrified by the press into some kind of infantile hot-box, a Skinner box for babies. It was simply an enclosed, ventilated and heated crib. The baby could look out but was somewhat insulated from the noise in the room. Unnecessary, of course – as I learned much later, babies need a little background noise, preferably muffled voices, to sleep well. After all, in our primeval past, complete silence meant the parents are gone and the wolves are on their way. The AirCrib was a passing fad, and we passed ours on to an eager Skinnerian when Nick – now known as "Woozle" after the imaginary Winnie-the-Pooh character – grew out of it.

Baby Nick in AirCrib.

My psychology office was on the fifth floor of the St. George Street building; my lab was in the basement. The main Psychology office, and most of the faculty were on the fourth floor. The two new hires on the fifth, Rolf Kroger and myself, were privileged: we each had a telephone. Nothing personal – the boon was granted us just because we were not close to the hub on the fourth floor. The peons on four (the five young faculty hired the previous year) had to use telephones in little cubbyholes in the corridor. I was happy for the phone. Easy access to a phone was pretty important for someone setting up a lab from scratch in pre-email days. I slowly got used to teaching and soon had a pigeon operant lab. I wrote a grant application – to the National Institute of Mental Health (NIMH) in the U.S. It was just 10 double-spaced pages long. I got the money. Now grant applications are 50 or so single-spaced pages, and you probably won't get the money – at least for the first three or four submissions.

At Toronto I had my one and only experience with an animal-rights agitator. Subsequently, I have had many, many harassing experiences with the animal-care bureaucracy. The two 'compliance' problems that stick in my mind are first, the requirement that we switch from wooden to metal cages for the pigeons (that happened at Duke University a few years later). And second, a citation for "mixed species" (nature is apparently in violation of that one), the two species in my case being crayfish (in a tank) and a bird (in a cage) in the same room. Go figure…Within a couple of weeks, two of our birds in the metal cages died. My pigeons didn't like metal cages. Racing pigeons, some worth thousands, don't either; they are also kept in wooden cages.

But Toronto provided my only encounter with an actual opponent of animal research. In any event this incident was mild, even humorous. One of the new faculty had a wife with a rather theatrical disposition. She clearly wanted to demonstrate against *something* and, for lack of anything better in super-bland Canada, chose our animal lab. She staged a sit-in. She sat, we came and went and, eventually, she departed. The pigeons, slim but warm and, as far as I could tell, perfectly happy, were unmoved.

On the intellectual side, Toronto was uncertain ground. In those days the lively field of animal learning was divided into two camps: Hullian and Skinnerian. The neo-Hullians, as they called themselves, were intellectual descendants of Yale's Clark Hull and his student Kenneth Spence. They were interested in the process of learning, not just behavior; they proposed theories; they studied groups rather than individual animals; and they used, of course, statistics. Indeed, they insisted on statistics even when they were completely unnecessary, as in individual-organism studies – this is one reason the Skinnerians started *The Journal of the Experimental Analysis of Behavior* in 1958. (If you need statistics, it usually just means you have a weak method – you can't easily replicate your result.) Groups were thought to be necessary because you cannot compare the effects of different experimental histories in the same animal. For example, suppose you want to see if an animal will learn to discriminate between two stimuli more quickly if trials are closely or widely spaced. Obviously, you can't do this with the same animal, you must compare a widely-spaced group with a closely-spaced one. You hope that the two groups are effectively identical, that the results of the group also hold for the individual and that statistical significance means the results can be replicated. The last two assumptions are often violated.

The dominant figure in Toronto at that time was Abram Amsel, a student of Kenneth Spence's at Iowa. Amsel was a neo-Hullian. Hullians, neo or otherwise, did not welcome either the criticism or the single-subject methodology of B. F. Skinner. Though far from Skinnerian myself, since I came from Skinner's lab I was definitely the odd man out. Since I had not given a talk, and my only publication at that time was on reinforcement schedules, the fact that I was a closet theorist was not widely known. I gathered from later conversations with graduate students that everyone was shocked and amazed at my appointment and expected that I would not last long. But personal relations were good and I sensed no hostility.

Frustration!

Trouble arrived when I had a bright idea about *frustration theory*, on which Amsel had built his entire career. Frustration theory conforms to everyone's subjective experience. But it has a conceptual flaw at its heart and a simpler alternative soon presented itself.

The theory rests on an intuitively plausible idea: that failing to get a reward when you expect it 'energizes behavior'. The door won't open to your first push, so you push it harder. The animal gets frustrated and consequently runs or presses faster – does more of whatever response he's been using to get the food. But beware explanations from introspection...

Frustration theory was tested initially with rats running in a 'double runway' that is, a runway with two goal boxes: one a short way from the start box and the second some distance after that: a short first runway (3.5 ft.) followed by a much longer second one (10 ft.). In training the rat runs to the first goal box and gets a bit of food, then he runs to second goalbox and gets another bit of food. The experimenter measures how fast the rat runs in the first part of the long second runway.

The experiment is in two phases. The rat always gets food in the second goal box (I'll call I the *endbox*). In the first phase, he also gets food in the first goal box (the *midbox*) on every trial. Thus, he learns to expect food in the midbox. In the second phase, he gets food in the mid box on only half the trials: rewarded on half, 'frustrated' on half. The question: How fast does he run in the long runway after food and after no-food, in the midbox?

The answer is: after training he runs faster when there is no food in the midbox compared to when there is food, especially in the first third of the long runway. This is the *frustration effect* (FE). Amsel attributed the effect to similarity. In the first phase the rat learned to associate the midbox with food, yet now (on half the trials) suddenly there is no food, hence 'frustration'. The more similar a given situation is to one where food is expected, the greater the frustration when food is omitted – hence the more vigorous the response.

There is a conceptual problem with this idea. It implies that frustration increases as the similarity of the no-food situation to the food situation increases. But the two situations are of course maximally similar when there actually *is* food. So should frustration be maximal when the rat actually gets food? Well, obviously not. Suppose the amount of food is smaller than usual: will that produce more frustration than complete absence of food? What does the theory say about that? There must be some kind of discontinuity: more similarity between food and no-food situation means more frustration – up to a point. This problem can be overcome with an additional assumption, but the fact that there is a problem obviously raises a question about the idea. Is there a simpler approach?

I had been studying timing behavior on fixed-interval reinforcement schedules. Time is not directly involved in the runway procedure, but I knew two things that made the FI studies relevant. First, any predictable delay between a time marker and food has an effect. On fixed-ratio schedules, for example, rats and pigeons pause after each food delivery even though pausing delays the next food unnecessarily. This is an automatic reaction to the minimum time between successive food deliveries imposed by the few seconds it takes to complete the fixed ratio requirement.

The long second runway in the frustration studies also imposes a delay between the food in the midbox and food at the end, in the endbox: the time it takes to run down the runway. The rats should therefore hesitate – run more slowly – right after midbox food. And they should hesitate more if the second runaway is long. Perhaps that's why the second runway is made much longer than the first? Perhaps that's why the frustration effect is largest in the first part of the second runway? Is the double runaway another example of experimenters unconsciously shaping apparatus and procedure so as to produce not so much a specific result, as a kind of result that fits preconceptions?

The second thing I knew from the interval-timing studies, is that food delivery overshadows any other available cue as a time marker. In the first phase of the experiment, when there is always food in the midbox, the rat's hesitation in the second runway is almost certain to be controlled by food, acting as a time marker, rather than by the other stimuli associated with his stay in the mid-goal box. And our work with fixed-interval schedules, suggested that his is likely to be true even though the food and the mid-goal box have the same temporal and spatial relation to the second food delivery at the end of the long runway. So, if food is in fact the effective time marker, then omitting the food should eliminate the hesitation. The rats should run faster, especially in the early part of the second runway, when the time marker – food – is omitted. Which is just what Amsel and his students had found.

So I had a simple alternative to the frustration hypothesis. Responding after food omission is not energized; rather responding after food is depressed, but after no-food, not so much. The frustration effect is an example not of excitation but of *disinhibition*.

Disinhibition is conceptually simpler than frustration, so it is perhaps not surprising that it had in fact been suggested earlier – by Alan Wagner, another student of Spence's and soon to become a major figure in associative learning. In his MA thesis in 1959 Wagner appeared to have disproved disinhibition by comparing a group that never got food in the midbox with nonreinforced trials with nonreinforced trials for the group that got food 50% of the time. As frustration theory predicted, the rats in the never-reinforced group ran slower in the second runway than the 'frustrated' rats on non-reinforced trials. It appeared that nonreinforcement was energizing only for the group that had learned to expect food. This result seemed to ditch disinhibition as an explanation and prove the frustration account. Fortunately, I was not aware of Wagner's apparently conclusive disproof of my idea until a year or two after I had tested it – fortunately, otherwise I might not have done the experiment. I'll explain why Wagner's experiment does not disprove disinhibition in a moment.

It was trivial to test the disinhibition idea with fixed-interval schedules. No need for a group of animals, one or two would suffice. So Nancy Innis and I designed an FI analog to the double-runway: a cycle of fixed-intervals, even-numbered intervals (analogous to the second runway) would always be reinforced; odd-numbered intervals (the first runway) would be reinforced only 50% of the time: instead of brief access to food, the lights in the box would go out for the same amount of time (called a timeout). Reward was 3-s access to food, delivered via a solenoid operated food hopper during a 3.2-s blackout; non-reward was the blackout alone.

Figure 8.1

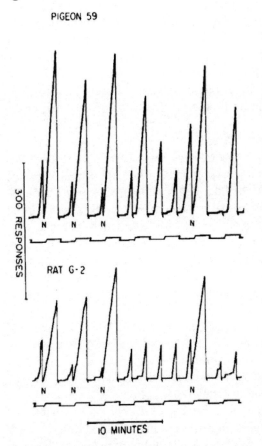

PIGEON 59

300 RESPONSES

RAT G-2

10 MINUTES

The Reinforcement-Omission Effect. Cumulative records of stable performance in an experiment in which a pigeon (top) or a rat (bottom) was trained on a fixed-interval 2-min schedule. At the end of intervals marked "N," food delivery was omitted and a brief stimulus was presented in its stead. The recorder pen was reset at the end of every interval. The number of responses (vertical scale) after non-reward ("N") is always much greater than after reward ("R"). (From Staddon, J. E. R., & Innis, N. K. (1969). Reinforcement omission on fixed-interval schedules. *Journal of the Experimental Analysis of Behavior*, 12, 689-700. Original study: Staddon, J. E. R., & Innis, N. K. (1966). An effect analogous to "frustration" on interval reinforcement schedules. *Psychonomic Science*, 4, 287-288.)

Prediction: a much shorter wait time after the timeouts than after food. The results (Figure 8.1) were exactly as we had expected – and much larger than the runway FE – which we also expected because pause induced by a 2-min fixed-interval schedule is much longer than the hesitation induced by a 10-ft runway.

With a naïveté that beggars belief, I presented our results to Amsel's group, fully expecting they would be delighted at clear proof of this new much simpler hypothesis. Er, no, not quite. I don't believe that Abe Amsel ever gave up his idea – although I heard from Bob Bolles, one of my teachers at Hollins, after he had moved to the University of Washington, in 1973 – that "Amsel is on the brink of giving it all up now." Bob added "It's about time I reckon." But Abe moved over into developmental psychology, studying infant rats, where frustration theory was reborn. It came to life a third time with research linking the frustration effect to regions of the brain.

Undeterred, I continued over the years intermittently to study the *reinforcement omission effect* (ROE), as it was now labeled. The essence of the temporal control (TC) idea is that the ROE depends on the behavior controlled by food, acting as a time marker. Several tests suggested themselves. Similarity between reward and non-reward figure both in frustration theory and in the temporal-control idea, but the predicted effects are opposite. For the FE, the more similar the non-reward situation to the reward situation, the more the frustration hence the larger the response. For TC, the more similar non-reward to reward, the smaller the elevation in responding.

John Kello, one of my students at Duke a few years later, suggested we do an obvious experiment, so obvious that I hardly thought it worth the effort, since it was sure to work as predicted. The experiment used the standard 2-min fixed-interval schedule. Kello varied the events produced at the end of each two minutes to make the complex more or less similar to the usual food reward. The events were: reward, which included food reward (R) plus the light illuminating the food magazine (ML), plus a 3-s blackout (B); no food, but B+ML; B alone; or no event (0). Intervals ending with one of these four events occurred in random order through each experimental session. We looked at the number of responses in the intervals following each.

The results were exactly as expected: the more similar the terminal event to the usual reward situation, the smaller the number of responses in the ensuing two minutes. The order (number of responses in two minutes) was 0 > B > B+ML > R+B+ML, precisely as the temporal-control idea predicted and the opposite of the frustration prediction.

The TC idea is that these omission effects all depend on the pattern of behavior controlled (temporally) by food. What if we trained pigeons with a schedule the opposite of the usual fixed interval, a schedule where they responded at a high rate after food and then after a while ceased to respond at all? Then omitting reward should have the opposite effect:

since reward produces a high response rate, non-reward should procedure a lower one. We designed such a schedule, the pigeons soon learned it and non-reward had the predicted effect, a 'negative frustration effect'. In short, every test we tried confirmed the TC view.

A few years after Toronto I used what is called a multiple schedule to test the idea that food reward overshadows a neutral stimulus as a time marker. A multiple schedule is just two (or more) different schedules occurring successively in the same experimental session, each signaled by its own stimulus (we've already encountered multiple schedules in the discussion of contrast in Chapter 6). The procedure worked like this: the basic schedule was again fixed-interval two minutes. The stimulus on the key was either red or green. The red intervals always began with N, the non-reward timeout (like Wagner's never-fed-in-the-midbox group). The green intervals began 50:50 with either N or R, the reward (like Wagner's 'frustration' group. Red and green intervals alternated at random. Thus in red intervals the time marker was always N, hence no possibility of interference[2] from R. But in the green, half the intervals began with N half with R so, according to the overshadowing idea, the wait time after N in those intervals should be shorter (and the number of responses therefore greater) than following N in red intervals. And so it proved. When there is the possibility of competition between N and R for control of the wait, R wins; but with no possibility of competition, N can serve as an effective time marker.

This is the real explanation for Wagner's apparent disproof of disinhibition. For the group never rewarded in the midbox, 'neutral' midbox cues are never overshadowed by intercalated midbox-food trials. Midbox cues can therefore exert good inhibitory control over running in the long second runway. The neutral stimuli acquire good temporal control – just like the stimuli in the red intervals in the multiple-schedule pigeon experiment. Running rate in the second runway will therefore be low – lower than on non-reinforced trails with the 50% group where food is the effective inhibitor and overshadows the midbox cues – like the green intervals in the pigeon experiment.

I regret that we didn't repeat some of these experiments with ratio schedules, which for some people are probably a more convincing analog to the runway. I would be amazed if the results don't confirm the TC hypothesis; but still, it would be nice to see. One simple possibility is a

---

[2] The stimulus on the pecking key in these procedures defines the situation, the animal's 'state', so that two 'operants' learned in the presence of two stimuli generally do not interfere with one another. The experiment I discuss here is described in more detail in Staddon, J. E. R. (1974). Temporal control, attention and memory. *Psychological Review, 81*, 375-391.

multiple schedule which alternates a small fixed ratio, 10 say, in the presence of a red key light, with a longer one, FR 30, signaled by green. This is analogous to the double runway. Then omission of 50% of rewards at the end of the FR10 should, as before, lead to a shorter pause, hence faster response rate in the FR30. But now, suppose we change the FR 30 to a random ratio (RR) 30. There is no pause once an animal has learned the RR schedule. So nonreward should have no effect: response rate after R and N should be the same. (Frustration theory would presumably make the same prediction as before: increased responding after N.) This would be another test of the basic idea: that these effects reflect the type of temporal control exerted by reward.

Maybe there is a real frustration effect. An experiment like this might reveal a trace of one. But if so, it is very much smaller than the disinhibitory process that we found. Quite possibly, like so much research that rests on 'null hypothesis statistical testing' (NHST), the frustration effect is simply an artifact[3] and doesn't exist at all. More on NHST later.

The failure of frustration theory does not mean that intuition doesn't exist. In the early evening of November 9, 1965, I was attending a late-afternoon seminar on an upper floor of the Psychology building on St. George Street. It was dark outside. The room had a wall of glass and all the lights of the city could be seen. When the meeting ended, I was the last to leave and I turned out the lights. At almost the same instant, all the lights in the city below went out. The subjective impression that I had caused the Great Northeastern Blackout of 1965 (it extended from New York to Canada) was quite overwhelming. The Law of Effect (temporal contiguity clause) in action!

Toronto isn't a bad city. It had its interesting neighborhoods. It did seem very empty, even compared to Boston much less Europe. I remember my first visit to the big Eaton's and Simpson's department stores downtown and wondering at the lack of people. "How do they stay in business?" I thought. (Amusingly, the apostrophes were omitted from these two names in 1972 because they might offend French-Canadians – although why the latter should be upset by apostrophes is something only a real Canadian can understand.) Not a bad city, but the winters are interminable and behind it all is, or was, a kind of reactive quality. The unspoken theme of Canadians seemed to be "We are not American! We are not American!" To which my equally unspoken response was: "I don't

---

[3] Research in the past decade has revealed the total inadequacy of many widely accepted statistical practices. See, for example, Pashler, H., & Wagenmakers, E.-J. (2012). Editors' introduction to the special section on replicability in psychological science: A crisis of confidence? *Perspectives on Psychological Science*, 7, 528–530 and a review in *The New Behaviorism*, 2nd edition Chapter 9.

care! I don't care!" Probably Canadians are more self-confident these days, after weathering the 2008 financial crisis much better than the U.S. and U.K. (the Bank of England now has a Canadian governor).

But the weather was the killer, and I was fond of the U.S. South from my very first experience of it. So I looked around. Duke University in North Carolina was advertising an assistant professor position, so I applied and, after a while, was invited for an interview. It was in January and Toronto was, as it usually is in the winter, cloudy and freezing cold. The chairman at Duke, Ned Jones, a charming social psychologist, drove me around sunny Durham in his powerful Oldsmobile 442 convertible, with the top down. It was one of those balmy interludes Durham often gets in mid-winter. At that point I would have accepted the job just for bed and board.

I gave a talk which went pretty well and had good interviews with several faculty. The person closest to me was Norman Guttman, a student of Skinner's when he was at the University of Minnesota. Norm had made important experimental contributions with his studies of generalization. Generalization, the fact that a response learned to one stimulus will still occur, albeit at a reduced strength, to similar stimuli, is an old idea. The great Ivan Pavlov was perhaps the first to propose that there is a gradient surrounding the training stimulus, falling off like the slopes of a mountain on either side. But until Guttman's experiment with Kalish, no one had found an easy way to demonstrate such a gradient. The persistence of responding, even in the absence of reward, by animals long-trained on variable-interval schedules provided the key. Guttman was able to show beautiful gradients in individual pigeons. Trained on a green spectral light, at 550 nm, for example, the pigeon's response rate fell off gradually in the presence of wavelengths on either side.

Norm was also very interested in the philosophy of psychology and had written much on the topic. I think he was quite taken by an in-press paper of mine on Skinner's concept of the *operant*. The operant is one of Skinner's more important theoretical contributions. The idea is that when an animal learns to get reward according to a given schedule in the presence of a given stimulus situation, that situation comes to control a sort of module or subroutine (not that Skinner ever used those words) with three components: response, reinforcer and stimulus. He called this concept a 'three-term contingency'. Each operant was 'controlled' by whatever stimulus signaled a given 3-term contingency. In my terms, an operant corresponds to a *state* of the notional organism. Re-reading my *Psychological Review* paper now, I'm afraid I find it quite mind-numbing – not just the rather abstract content, but the required passive voice and lumbering academic prose. Nevertheless, Norm was keen on epistemology, my empirical research was pretty interesting and so I got the job.

Before we left Canada, in the summer of 1967, we went, all three, to Montreal for the Canadian World's Fair: "Expo 67". I had a good friend in Toronto, another one of the cohort of young faculty. Bernie Schiff had done his PhD at UC Berkeley with Steve Glickman, a world expert on the behavioral biology of spotted hyenas. Glickman set up a hyena colony, a unique project, that persisted until it was finally shut down in 2014 for lack of funding. Hyenas are fascinating animals, with a female-dominated society (they are bigger than the males) and very odd genitalia. The female genitalia resemble the male, and they even have a sort of clitoral 'pseudo penis': they are the LGBT champs of the animal kingdom. Laurence Frank, co-founder of the colony, found the young animals he brought back from Kenya to be "great fun" and "more tameable than wolves".

Bernie's family lived in Montreal and they very kindly put us up for a couple of days that turned out to be more exciting than we could have anticipated. Because it was during our visit that the Israel-Arab Six-Day War (June 5-10, 1967) began. Bernie had relatives in Israel and there was high anxiety all round at the prospect of a war between tiny Israel and its populous Arab neighbors. But Israel, seeing opposition forces massing on its borders, acted pre-emptively and the war was short – amazingly so, considering the apparently interminable Middle-Eastern conflicts that have been going on for the past decade and a half. Israel has never been forgiven by many in the West for her success and the huge disproportion in casualties – perhaps 1000 Israeli lives to twenty or more thousand Arab. But what was the alternative in the face of an existential threat?

Bernie and I used to talk at length in Toronto – to the point that my wife sometimes felt neglected and became I think even a little jealous. But Bernie's scientific interests drifted from his Glickman origins and we lost touch after I moved to Duke.

So we visited Expo 67, much enjoyed by Woozle, and admired all the new construction, especially the odd, Lego-like apartment buildings designed by Canadian architect Moshe Safdie. But the South beckoned.

Chapter 9: North Carolina and 'Superstition'

So began an association with Duke University that lasts to this day. Duke evolved from pre-civil-war Trinity College, a school for Methodist preachers. The Duke motto, *eruditio et religio,* appears to give religion a major place in the mission of the university. But the Methodist identity has grown feeble over the years, replaced by an eclectic monotheism. So eclectic that in 2015 there was a serious suggestion that the Duke imam (yes, there is one) should be allowed to issue the Islamic Friday call to prayer from the tower of the Duke Chapel.

Duke has become one of the top half-dozen U.S. private universities. The first international student at Trinity, in 1881, a boy called Charlie Soon (later Soong), was a Methodist protégé. He was to become the influential patriarch of a family with connections to Sun Yat-sen, founder of the Chinese republic, and Chiang Kai-shek, the opposition to Mao Zedong during WW2 and first leader of an independent Taiwan. An improbable connection for a small southern college, but prophetic of Duke's international ambitions.

The big step in the evolution of the university was the creation of the Duke Endowment in 1924. Duke University was and is the main beneficiary of the endowment. James Buchanan "Buck" Duke, who had made his mega-money from tobacco – cigarettes were Durham's main industry for many years – was the chief benefactor. Buck Duke's largesse was eased by a suggestion that the far-from-unique college name 'Trinity' be replaced by 'Duke'. So in 1922, Duke University was born[1]. A splendid neo-Gothic campus a mile distant from the original Trinity College arose in the 1930s, overlooked by a large and impressive chapel. The architect for the West Campus (the old college is the East Campus) was the Philadelphia firm Horace Trumbauer. In 1986 diligent digging by socially conscious students revealed that the actual designer was Julian Abele, the African American chief designer and draftsman of the firm – a fact long known but insufficiently proclaimed. Much has been made of this Duke link to talented black America in the years since.

Duke's race history is in fact pretty good. A minor incident in 1903 involving segregation and Booker T. Washington – now a hero, but then not, at least in the South – caused a Trinity College faculty member, John Spencer Bassett, to call Booker T. "the greatest man, save General Lee, born in the South in one hundred years." This outraged many. The college

---

[1] John K. Winkler (1942) *Tobacco Tycoon: The story of James Buchanan Duke.* New York, Random House.

resisted loud calls – from the proprietor of the now smugly progressive[2] (Raleigh) *News and Observer*, among others – for Bassett to be fired. (Nowadays it is more likely the reference to General Lee that would cause offense.)

Tobacco was celebrated when I first came to Durham. The Liggett and Myers factory hosted tours which ended with a free pack of Luckies. The aroma of tobacco pervaded the town. All the cigarette factories are now gone and the sturdy brick tobacco warehouses are converted to other uses, from art centers to cafes, mini-malls and Duke annexes. Contemporary Duke is embarrassed by its links to the fatal weed, bans smoking almost everywhere on campus and welcomed a Medical Center initiative to label Durham 'City of Medicine'(C of M). They hoped, perhaps, to erase its previous title: 'Bull Durham' – named after a world-famous smoking tobacco. The movement failed all but financially. The local baseball team is the Durham Bulls; a large bronze bull adorns the central square and 'Bull Durham' (starring Kevin Costner, Susan Sarandon) makes a better movie title than *C of M*. But healthcare is the biggest employer in Durham. Somehow, it was more fun when tobacco was king.

Three graduate students came with me from Toronto when I set up at Duke. One was Nancy Innis; the second was Janice Frank, who was writing up her Toronto PhD dissertation and in the meantime worked as my lab assistant. The third was a young woman called Virginia Simmelhag. Ginny was finishing her Toronto MA thesis. She had been an undergraduate in the 'General' (as opposed to the 'Honours' program in Toronto) – i.e. she was classified as not-so-good. Phooey!... she was one of the best young scientists I ever hope to encounter. A wonderfully acute and balanced observer, she had done a "what if?" experiment based on an influential 1948 study by B. F. Skinner. She departed later with her husband to Memorial University in Newfoundland, where after some years she completed her PhD with Sam Revusky, a creative experimenter and, unusually for a psychologist, also a rabbi.

Theories of learning *are* necessary

Skinner's experiment was unbelievably simple. The procedure was Pavlovian rather than operant. That's to say: no response was required from the pigeon, who got brief access to food every 15 s independently of his behavior. The experiment was novel in another way: the pigeon's behavior was observed, not recorded automatically, unlike almost all of Skinner's other studies.

---

[2] The N & O was progressive then as now, but editor Josephus Daniels also advocated white supremacy.

To understand the significance of this experiment I need to give a little background. By the 1940s, learning psychologists recognized two kinds of conditioning: *operant* – all the experiments I've discussed so far are operant conditioning – and Pavlovian or *classical* conditioning. Skinner had argued – persuasively but, I think, in the end wrongly – that these two procedures are fundamentally different. Operant conditioning, training a response by reward or punishment, requires a dependence of the reward on the response. This Skinner termed a *reinforcement contingency*, but the idea is familiar. If you want the dog to lift his paw when you say "paw!" you give him a treat when he does so. The treat is contingent (dependent) on the movement. Behavior that is responsive to reinforcement contingencies like this is termed 'operant'.

In Pavlovian conditioning there is no dependence on behavior; there *is* dependence of reward (or punishment) on a *signal*. A tone sounds, and Pavlov's dog gets a bit of food. No response required but, after a few such pairings, the dog will salivate at the sound of the tone. This is Pavlovian conditioning and the responses, like salivation, that are susceptible to it Skinner called *respondents*. Respondents were thought to be quite distinct from the kind of skeletal, 'voluntary' behavior that can be conditioned by consequences – *operants*.

The stimulus for classical conditioning need not be a tone or a light. *Time* – the fixed time between successive food deliveries, for example – can work as well. This is called *temporal conditioning* (or, in the Skinnerian jargon, a fixed-*time* schedule). Skinner's experiment in fact used temporal conditioning. Since no operant contingency is involved, no operant behavior should occur. But 'unnecessary' operant behavior, ranging from pacing to wing-flapping and 'incomplete pecking movements' did occur – at a high rate. Hence Skinner's title for the paper: "'Superstition' in the pigeon". So, what are we to make of the operant-respondent distinction and the various learning theories that have been built around it?

Ginny and I were not conscious of these theoretical issues when we designed her experiment. It arose as a result of two things. First, I had kept up with the growing ethological (animal-behavior) literature that involved careful observation of animals' behavior: second-by-second identification of specific behavioral categories. I thought it might be interesting to look at operant conditioning this way. Second was a strictly empirical question: Skinner had used a fixed-time schedule, so what would happen if we used a variable-time schedule; what if we gave the pigeon a bit of food not at fixed intervals of time but at more or less random times? Would we get superstitious behavior then? Ginny's experiment sought no grand theoretical breakthrough; it was curiosity, pure and simple.

So we compared fixed- and variable-time schedules. But Ginny coded the behavior very carefully. She pushed a button to identify just which behavior was occurring and these button pushes operated an event recorder so that the exact time of each behavior could be measured. We replicated Skinner's result and got superstitious behavior even on the variable schedule. But the details were very different.

Skinner's use of the term 'superstitious' is another example of his rhetorical skill. It immediately caught on and allowed him to do what later became his métier: extrapolating from pigeons to people. He introduced the 1948 paper with this famous paragraph:

> To say that a reinforcement is contingent upon a response may mean nothing more than that it follows the response. It may follow because of some mechanical connection or because of the mediation of another organism; but conditioning takes place presumably because of the temporal relation only, expressed in terms of the order and proximity of response and reinforcement. Whenever we present a state of affairs which is known to be reinforcing at a given drive, we must suppose that conditioning takes place, even though we have paid no attention to the behavior of the organism in making the presentation[3].

He concluded "A simple experiment demonstrates this to be the case." So any operant behavior that occurs in this situation is to be explained by 'adventitious reinforcement', accidental contiguity between behavior and the delivery of food. No mention of the apparent contradiction: operant behavior induced by a Pavlovian procedure. Skinner evaded that issue by providing up front a clever alternative account: operant conditioning always occurs, even when reward is independent of any response. Nothing to see here – superstitious behavior is not a problem.

The 'superstition' paper is the only one in Skinner's abundant *oeuvre* that proposes to test a hypothesis, a practice he roundly criticized in an article published a couple of years later called 'Are theories of learning necessary?' ("No" was his answer.) Why was the 'superstition' paper Skinner's only hypothetico-deductive one? Thomas Huxley famously spoke of the tragedy of science as "the slaying of a beautiful hypothesis by an ugly fact." Skinner managed to obscure an obvious contradiction by a beautiful argument.

Ginny's pigeons showed superstitious behavior, but it did not fit Skinner's account. When our fixed-time 12-s procedure began, the

---

[3] Skinner, B. F. (1948) 'Superstition' in the pigeon. *Journal of Experimental Psychology, 38,* 168-172.

pigeons did show a behavior that reliably preceded food delivery: they always had their heads in the feeder opening. That is the behavior which should have been conditioned, according to Skinner's account. But it was not. Instead pecking, or a behavior close to pecking, appeared spontaneously after a few sessions. We called this the *terminal response.* Rather than being variable in form, as Skinner's account implied, it was in fact quite stereotyped: pecking or something close to it. The pecking we observed occurred full-blown, with no history of accidental contiguity with food.

Skinner's account also failed to fit a second category of behavior that we called *interim*: wing-flapping, jumping, pacing are examples. These activities are quite variable from bird to bird. They occurred earlier in the interfood interval so they were rarely contiguous with the food. 'Adventitious reinforcement' could not explain them either.

In fact, we had replicated with temporal rather than standard Pavlovian conditioning, a result published a year or so after we did the experiment in Toronto. That experiment was a demonstration by Herb Jenkins, a talented Skinner student now a professor at McMaster University in Canada, of something he and his co-author called *autoshaping.* Jenkins used the more familiar Pavlovian procedure, not temporal conditioning but standard delay conditioning. Every minute or so the response key turns red for 7 s and the food is delivered. After a few such pairings, 'naïve' pigeons peck the red key as soon as it appears. Jenkins presented this result, so contrary, really, to accepted ideas about the roles of operants and respondents, simply as a handy automated way to train naïve pigeons to key-peck.

Even the sequential properties of the behavior that Ginny recorded so carefully contradicted popular theory. Despite an influential 1951 paper by the physiological psychologist Karl Lashley – 'The problem of serial order in behavior' –most behaviorists still thought of behavior sequences as stimulus-response chains, each behavior in a sequence providing the stimulus for the next. But in a chain, ...ABC... say, omission of B must prevent the occurrence of C. Yet the activities in the sequences we saw seemed to be linked to time – postfood time – not the behavior that usually preceded them. Sequences containing AC or just C alone were perfectly possible. Subsequent experiments clearly showed that chaining is a myth. We found, for example, that the later in a fixed interval the pigeon began to peck, the faster he pecked[4]. No way to fit that into a chaining account; post-food time is the real driver.

---

[4] Staddon, J. E. R., & Frank, J. A. (1975). Temporal control on periodic schedules: Fine structure. *Bulletin of the Psychonomic Society*, 6(5), 536-538. http://hdl.handle.net/10161/5997

The idea that the 'superstitious' terminal response is maintained in some way by its (admittedly accidental) consequence was conclusively disproved a few years later, when two scientists at the University of Pennsylvania showed that an autoshaped pigeon will not cease to peck even if a peck on the light turns it off, actually preventing food delivery on that trial.

When it came time to write up our observations I had to consider seriously the very many ways in which they contradicted standard accounts, Skinner's in particular. What seemed to be important was not contiguity with food but being in a situation that strongly *predicts* food – like the end of the 12-s FT or Jenkins' 7-s conditioned stimulus. In the mid-1960s, two students of Skinner's, Marion and Keller Breland trained animals like chickens and raccoons for use in advertising. They reported dramatic examples of 'misbehavior' cued by 'hot' stimuli. For example, a raccoon can be trained to pick up a token and deposit it in a piggy bank which then dispenses food. But not for long. Once the 'coon has learned the close connection between token and food, he spends more and more time 'washing' the token, thus delaying and finally, when he won't quit, preventing food delivery. The Brelands called these odd behaviors 'instinctive drift' and reported them in a famous article. But they did not see them as anything more than interesting exceptions to the near omnipotence of operant conditioning: "[Skinner's] point was that what is seen as misbehavior is simply a failure to control, manipulate, or at least recognize all of the variables. That argument we concur with..." said Marion many years later[5].

Evidence contrary to Skinner's account was piling up. So we submitted a moderately theoretical article describing the experiment to the *Psychological Review*, the premier theoretical journal in psychology. After getting good reviews, the editor, George Mandler, gave us a response quite unique in my experience. He suggested that we *expand* the theoretical discussion, which we did. The result was a long article that became a 'Citation Classic' and a blueprint for much of my future research[6].

The picture that began to emerge then and has come increasingly into focus over the years is a Darwinian one. Lacking any course in biology, I came to Darwin rather late. I do remember in Richmond during my last college year, looking into the *Origin* and reading a few articles about evolution. Once I understood evolution by natural selection, then, like

---

[5] *The Psychological Record*, 2010, 60, 505–522.
[6] Staddon, J. E. R., & Simmelhag, V. (1971).The "superstition" experiment: A reexamination of its implications for the principles of adaptive behavior. *Psychological Review*, 78, 3-43.

Thomas Huxley, I also thought "How extremely stupid not to have thought of that!?" I subsequently read almost every one of Darwin's major works: *The Voyage of the Beagle, Coral Reefs, The Variation of Animals and Plants under Domestication, The movements and habits of climbing plants, The Descent of Man* and of course his book on psychology *The Expression of the Emotions in Man and Animals*. I skipped the treatise on barnacles, although it is still, I understand, the definitive work on these creatures. I was also familiar with a number of other attempts to relate learning and Darwinian ideas – by the Oxford zoologist J.W.S. Pringle, and psychologists such as D. T. Campbell and Skinner himself.

Figure 9.1 From *Coral Reefs*, to illustrate the process of formation.

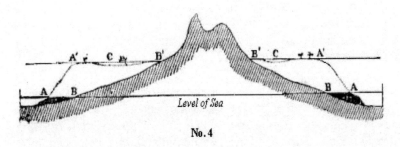

No. 4

AA—Outer edge of the reef at the level of the sea.
BB—Shores of the island.
A'A'—Outer edge of the reef, after its upward growth during a period of subsidence.
CC—The lagoon-channel between the reef and the shores of the now encircled land.
B'B'—The shores of the encircled island.

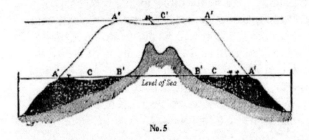

No. 5

A'A'—Outer edges of the barrier-reef at the level of the sea. The cocoa-nut trees represent coral-islets formed on the reef.
CC—The lagoon-channel.
B'B'—The shores of the island, generally formed of low alluvial land and of coral detritus from the lagoon-channel.
A"A"—The outer edges of the reef now forming an atoll.
C'—The lagoon of the newly formed atoll. According to the scale, the depth of the lagoon and of the lagoon-channel is exaggerated.

Darwin's book on coral reefs shows how he thought. He had the germ of an explanation before he even saw a reef, mentioning it in the *Beagle* book. His first creative step was to group together the three main kinds of coral reefs: Fringing reefs are directly attached to a shore, or border it with an intervening shallow lagoon. Barrier reefs are separated from a mainland or island by a shallow lagoon. Most fascinating, and puzzling, are atolls, which are circular or C-shaped with no central island.

How did these strange structures, varied but subtly similar, come about? Coral reefs are made by small polyps, *Cnidaria,* that create an exoskeleton, a protective structure that is basically chalk: calcium carbonate. Colonies of them over time build a massive rock structure, the reef. The puzzle – this part was not known to Darwin – is that the coral-created rock may go down hundreds of fathoms beneath the waves, even though coral polyps cannot live in the depths. The commonest type lives in symbiosis with photosynthetic organisms that can survive only in shallow clear water.

In Darwin's day, geographers assumed that reefs built up on just-submerged surfaces, or were formed on the sunken craters of volcanoes. But there were problems with these accounts. He disposed of them in typical Darwin fashion by listing overwhelming contrary data:

> The naturalists who have visited the Pacific, seem to have had their attention riveted by the lagoon islands, or atolls,—those singular rings of corall and which rise abruptly out of the unfathomable ocean—and have passed over, almost unnoticed, the scarcely less wonderful encircling barrier reefs. The theory most generally received on the formation of atolls, is that they are based on submarine craters; but where can we find a crater of the shape of Bow atoll, which is five times as long as it is broad.... or like that of Menchikoff Island ... with its three loops, together sixty miles in length; or like Rimsky Korsacoff, narrow, crooked, and fifty four miles long; or like the northern Maldiva atolls, made up of numerous ring formed reefs, placed on the margin of a disc,— one of which discs is eighty eight miles in length, and only from ten to twenty in breadth? It is, also, not a little improbable, that there should have existed as many craters of immense size crowded together beneath the sea, as there are now in some parts atolls. But this theory lies under a greater difficulty, as will be evident, when we consider on what foundations the atolls of the larger archipelagoes rest: nevertheless, if the rim of a crater afforded a basis at the proper depth, I am far from denying that a reef like a perfectly characterised atoll might not be formed; some such, perhaps, now exist; but I cannot believe in the possibility of the greater number having thus originated.

And why the lagoon? And why so shallow? Why do reefs occur only in certain areas, parts of the Pacific, especially? In his autobiography Darwin writes:

> No other work of mine was begun in so deductive a spirit as this; for the whole theory was thought out on the west coast of S. America before I had seen a true coral reef...I had therefore only to verify and extend my views by a careful examination of living reefs. But it should be observed that I had during the two previous years been incessantly attending to the effects on the shores of S. America of the intermittent elevation of the land, together with denudation and the deposition of sediment. This necessarily led me to reflect much on the effects of subsidence, and it was easy to replace in imagination the continued deposition of sediment by the upward growth of coral. To do this was to form my theory of the formation of barrier-reefs and atolls.

Darwin had noticed evidence of subsidence in the Pacific. He knew that the coral polyps like living in shallow water with breaking sea. He put these two things together in a model in which two opposing tendencies competed to produce the result: upward growth of the coral opposing downward descent of the land. If the coral can grow faster than the land sinks, it will build up and up around what was once a shoreline to create the three kinds of reefs. Only in the 1950s was deep drilling applied to show that indeed, as the theory predicts, below an atoll the coral-created chalk may go down many hundreds of feet – much deeper than a living coral can survive.

Darwin applied this line of argument – explaining a set of facts by the action of opposing processes – in several other areas, most famously in evolution by natural selection. But he was not averse to looking at trivial phenomena in the same way. In his *Life and Letters* we find the following account of ants carrying their cocoons:

> The ants carrying the cocoons did not appear to be emigrating... But when I looked closely I found that all the cocoons were empty cases... Now here I think we have one instinct in contest with another and mistaken one. The first instinct being to carry the empty cocoons out of the nest... And then came in the contest with the other very powerful instinct of preserving and carrying their cocoons as long as possible; and this they could not help doing although the cocoons were empty. According as the one or other instinct was the stronger in each individual ant, so did it carry the empty cocoon to a greater or less distance.

Thus, Darwin suggests, two opposed tendencies drive ants to remove their empty cocoons to a safe distance from the nest.

So to natural selection and operant learning. Operant conditioning is also the result of two opposing processes: selection and variation. The selection is done by reinforcement. The process is still not completely understood but the outlines are clear. Contiguity (does the reward closely follow the response?) and competition (does the reward occur at other times?) are involved. The sources of variation, the occurrence of behavior in advance of reinforcement – Skinner's 'emitted' behavior that occurs for 'other reasons' – were also becoming clearer. Our experiment and the many experiments on autoshaping showed that Pavlovian conditioning plays a major role in the 'variation' part of the dyad.

Cognitive terms can cause trouble because they are often derived from folk psychology, from introspection, which is a notoriously unreliable guide to the true causes of behavior. 'Memory' with its smorgasbord of types – long-term, short-term, 'working', 'episodic', 'declarative', 'procedural', etc. – is one term that continues to give problems. 'Expectation' is another. But Pavlov's identification of classical conditioning with the reflex activity of the autonomic nervous system – with salivation – has misled psychology in the opposite direction. If 'expectation' means too much, labeling conditioning as a reflex means much too little. The salivation of Pavlov's dogs is perhaps the least important effect of the conditioning process. What is important is indeed its effect on the animal's *expectation*, that is to say on his *repertoire*, the set of candidate activities he brings to this new situation – his *state* in my terms. If a stimulus signals food, then food-related activities are activated; if it signals a mate, then sexual activities; if a rival, then aggression, and so on.

These activities are available in advance of any consequence. They are *emitted*, to use Skinner's word, not elicited like a true reflex response. They occur in succession, strongest first. If the activation is not too strong, then if an activity is ineffective it will weaken to be supplanted by another. In this way an effective activity (if there is one within the animal's repertoire) will eventually be selected. The selection rules are such things as temporal proximity: did the reward happen right after the response? And predictiveness: did the reward occur at times other than right after a response? But if the activation remains strong – reward continues to occur, so predictiveness is high for any activity – then an irrelevant or even maladaptive activity may persist indefinitely. Hence 'superstitious behavior' and the terminal response, which occurs if the animal is very hungry (Skinner's pigeons were at 75% of their free-feeding weights) and food is given frequently. Superstitious behavior does not occur if the pigeon is not hungry or food only occurs (say) once a minute.

Activation can override consequence. For example, there is something called the 'feature-negative effect'. It works like this. The pigeon is trained to discriminate between two similar stimuli, say **° and ***. The unrewarded stimulus is distinguished by a feature, in this case the °: *** is the rewarded stimulus, **° is unrewarded. The two stimuli are perfectly discriminable, in the sense that under appropriate conditions, the pigeon can learn to peck either or both, depending on the payoff schedule. He can easily tell the two stimuli apart. But if the schedule yields rewards at a high rate, the pigeon may fail to behave differentially, responding equally to both stimuli. The tendency to peck, induced by a 'hot' situation, overrides the relatively weak selective effect of non-reward. On the other hand, if the stimuli alternate at, say 3-minute intervals and the positive one is paid off only once a minute on average, the birds will soon cease to respond to the negative stimulus. And if the contingencies are reversed, so that **° is the positive stimulus and *** the negative, the bird has no problem even if the reward rate is high: He pecks at °.

Skinner was not totally wrong about adventitious reinforcement, however. Many years later, Zhang Ying and I were able to show that a very simple contiguity-based model of selection (reinforcement) could produce something like superstitious behavior and even instinctive drift. This bare-bones learning model showed that a frequent 'free' reward yields behavior in long runs: AAAAAABBBBBB... rather than ABABBAB. Conversely, if rewards occur infrequently, the sequence of activities looks more random. So if the observer's sample is limited and rewards are frequent, the bird may seem to have fixated on one activity. Instinctive drift – rewarding A and then, eventually, getting B – happens if activity B 'conditions' more slowly than A but reaches a higher level. So Skinner's suggestion is theoretically feasible, although it's not what happens in practice[7].

Zhang Ying was a graduate student from China. She was very bright and a good collaborator, but I always sensed that research wasn't her real aim. So one day I had a heart-to-heart with her and asked her what she really wanted to do. Well, it turned out that her career was set by her parents. They wanted her to get a PhD and that was that. But what she really wanted to do was be a kindergarten teacher! Her parents had in fact arranged for her to marry just before she left China for the U.S. The

---

[7] Staddon, J. E. R. & Zhang, Y. (1991) On the assignment-of-credit problem in operant learning. In M. L. Commons, S. Grossberg, & J. E. R. Staddon (Eds.) *Neural networks of conditioning and action, the XIIth Harvard Symposium.* Hillsdale, NJ: Erlbaum Associates. Pp. 279-293, reprinted in 2016.

marriage had never been consummated and her husband remained in China for a while. But after a year or so, he reappeared, as a graduate student (in engineering, naturally) at NC State in Raleigh. They moved in together; Ying in due time had a child and apparently they lived happily ever after – or for as long as I was in touch, anyway.

Dear Sir:

I feel sorry that I was very aggressive to you that afternoon. I do apologize if it hurts your feeling although I didn't mean to do that. I think I just didn't know how to handle all the things especially myself those days. I was worried and anxious because I couldn't figuer out what's my research for, what's my study for and also what's my life for. Almost six weeks of thinking, now I know that life is a gift and pressure is a gift, maybe there are some reasons for their existence maybe not. But I should and I can face them and deal with them anyway.

I think I should thank you a lot because your are special in my life. You are the first person who taught my what the science was and how to do research work on science. You are the first person who told me that life here is a pain. And you told me many other things I didn't know or notice before. Thanks again for your help after I came here.

Today is your birthday. It's a good opportunity for me to say what I wanted to but what I didn't know where and how I could express with my best birth-day wishes. HAPPY BIRTHDAY TO YOU, SIR.

Sincerely yours, Ying
3/19/90

A rather touching handwritten letter I got from Chinese graduate student Zhang Ying in 1990 after a conversation that was obviously much more difficult for her than for me – I only remember her as always very sweet and kind. And, no, I never expected her to call me 'Sir'!

Ying also illustrated a well-known stereotype that was explained to me at the time by a Chinese colleague. Driving was then a great novelty to many new Chinese immigrants. Both space and physics caused problems. Chinese novices had a tendency not to slow down before turning (physics), and they tended to get lost (space). Ying eventually got a car, but then took a wrong turn after dropping a friend at Raleigh-Durham airport and was lost for some three hours (all this before GPS, of course). She eventually had a fender bender and gave up driving. No studies seem to have been done on this politically sensitive problem. It has probably disappeared now that China has upgraded from bicycles to cars.

Staddon lab, circa 1990, in front of the Psychology-Sociology Building at Duke *Left top:* **Nancy Innis (visitor), Vanessa Wells (secretary), JS, Kazuchika Manabe (visitor), Derick Davis.** *Front:* **Armando Machado, unknown undergrad, Mark Cleaveland, Jennifer Higa (postdoc).**

Skinner never abandoned his original hypothesis about superstitious behavior. As late as 1983, in the third volume of his autobiography, he wrote: "Give a hungry pigeon food every twenty seconds, and it will develop a superstitious ritual. The food reinforces whatever the bird is doing at the time, even though there is no "real" connection. A second instance is then more likely to occur, and the behavior may become so strong..." It's tough to give up a bright idea, especially if it is popular hit.

So much for science. Meanwhile, DB and I set up our new house – an elegant, modest-sized but in many ways impractical, flat-roofed and architect-designed (Arthur Cogswell) house with beautiful grained oak floors: 2719 McDowell Road, not too far from the Duke campus. It had been built only two years before. The owners were selling to move to presumably greener, but certainly cooler, academic pastures in Wisconsin. After a few years we added a wing and I built a bookcase wall and door to make another room on the ground floor. Then, the house was sold, the wall was removed; the house was sold again and another wing was added to the front – all within 25 years or so. Such is the transformative way with American suburban properties. The house still looks nice, though.

Durham is woods. The short winter and ample rainfall ensures that every square foot of ground, left to itself will fill with trees, beginning with loblolly pine and ending with mature deciduous forest of sycamore, oak, tulip poplar and sweet gum. Our new house sat on two lots, a total of almost an acre, backing on a strip of Duke-owned woods and a two-lane road. Ridges showed our acre had once been cultivated; but now it was a forest, mostly of pines. Quite a change from Toronto which was mostly treeless and, for most of the year, also leafless. We were surrounded by trees, to the point that a few years later I had to pay a local company to cut down six of them that were threatening to fall on the house. It prompted my only graduate-student-help party, a *Holzfest*. We provided pizza, a few of the more burly and kind-hearted of my students provided log-splitting.

Tree cutting is dangerous. One of the team of brothers who cut down our trees died in an accident just a few years later. Log-splitting also has its hazards, but in our case they went no further than aching backs.

One of the side-benefits of our move to NC was proximity to DB's father in Roanoke. DB and he were reconciled and he had become attached to toddler Woozle. But then, tragically, John Ballator died the year we arrived, suddenly at the very young age of 58. Born in Oregon and still with a considerable reputation there, he was head of the lively Hollins Art Department and a skilled representational artist himself. He did a large WPA painting in the Department of Justice building in Washington DC and several others in Portland. We inherited his car and a little money, which we spent, appropriately, on art.

In March 1964, during my last year in Cambridge, we had visited an art show at the De Cordova Museum in Lincoln. The artist was M. C. Escher – this may have been his first American show. We were impressed by the cleverness – the psychological cleverness – of his now-famous etchings and lithographs. I have always thought of Escher as perhaps the last in a line of great artists who were driven by both esthetics and

Staddon Lab alumni in 1993: Nancy Innis, JS, Mike Davis, John Kello, Janice Frank, John Malone, Ginny Simmelhag-Grant.

science. Leonardo is the obvious example, but all the classical artists were trying to understand and represent nature – truth – as best they could. Escher seemed to have the same motive, but he explored the psychology of perception. Anyway, we liked Escher. So when DB inherited a little money we wrote to him, in Baarn, in the Netherlands, and asked if we could buy some pictures. He was happy to do it. He identified the pictures by directing us to an illustrated book of his work. We bought in the end eleven or so. The most expensive, 'Ring Snakes' – his last work – cost us just $300.

A few years later, the flat roof of our McDowell Road house considerably diminished our art collection. The roof was badly built, with no air circulation between the insulation and the plywood of the roof above. In consequence, condensation formed and now the plywood was rotting. None of this was visible from outside, of course. So we had to replace the entire roof. The cost was several of the Eschers, which had risen considerably in value. The few that remained I donated many years later to the Nasher Museum at Duke. A local expert made four beautiful sets of copies – better in many respects than the original. Second wife Lucinda and I gave a set to each of our four kids.

Alas, the tradition that animated Escher seems almost extinct in the art world. Too many successful contemporary artists seem motivated more by gimmick, publicity and provocation than skill and science. I suspect that most of the real artists have migrated to advertising, album covers, CGI and video games. The art world has never really recovered from the invention of photography.

In 1969 our daughter Jessica was born. Even as a child she seemed mature. I have a picture of her, at age 11, in the National Anthropology Museum in Mexico City, looking at least 18 and completely in charge. This was not an illusion; she has always been a very self-contained and independent person. After finishing high-school she decamped to California to attend UC Berkeley, where she studied various kinds of mathematics ending with a PhD in combinatorics. She acquired a husband, Mike Barnes, and two boy children and made a very successful career in encryption and privacy. She returned (finally!) to NC in 2015 from a job at Google to teach at NC State in Raleigh.

Chapter 10: Time and Memory

In 1973 I published a long philosophical account updating the purple-paper submission to Skinner's course at Harvard. It was an attempt – largely unsuccessful at that time – to shift the center of gravity of behaviorism away from his 'no theory, behavior only' theme. The title "On the notion of cause, with applications to behaviorism"[1] was a nod to Bertrand Russell's famous paper in which he questioned the idea of cause and effect as a basis for science. I questioned the idea that stimulus and response are an adequate basis for psychology. I proposed 'state' as an essential third element. But now, more than 40 years later, the message seems to be getting through. I was recently asked by a behavior-analysis group to give a talk; not just any talk, but a talk on *models*.

My first years at Duke were devoted to experimental studies of timing behavior. The precision with which an individual pigeon can adjust its behavior to complex temporal contingencies is really extraordinary. For example, still showing traces of the matching-law influence, I did an experiment on choice, using spaced-responding schedules. The title of the paper is "Spaced responding and choice: a preliminary analysis"[2]. (We were more modest then. But it's a long paper: now it would probably be titled "Spaced responding and choice: a definitive analysis").

Timing, choice and power functions

The pigeon pecked a single key. He was paid off, on a variable-interval schedule (now a standard in choice experiments), for interresponse times (IRTs) between 2 and 3 s. He was also paid off for any longer IRT between 10 and 11 s. The result was beautiful double-humped IRT distributions, with peaks at 3 and 11 s. The trough was at 6 s. I varied the VI rate and waited until behavior stabilized at each rate. The two peaks varied reciprocally. As more rewards went to the shorter IRTs, so that peak increased and the longer one decreased. I counted the number of IRTs less than and greater than 6 s (short vs long), generating a table of long and short response rates vs reinforcements for long and short. A plot of the ratio of short/long responses vs reinforcers received for short and long was very orderly, and well fitted by a power function. This was the

---

[1] Staddon, J. E. R. (1973). On the notion of cause, with applications to behaviorism. *Behaviorism*, *1*, 25-63. Reprinted and translated as Sobre a noção de causa: Aplicações ao caso do behaviorismo. *Cadernos de História e Filosofia da Ciência*. 1981 (4) 48-92.

[2] Staddon, J. E. R. (1968). Spaced responding and choice: A preliminary analysis. *Journal of the Experimental Analysis of Behavior*, *11*, 669-682. http://dukespace.lib.duke.edu/dspace/handle/10161/5995

first time that operant choice data had been fitted by the power-function alternative to simple matching. (In the same journal issue, two antipodean researchers suggested the same equation to describe successive discrimination performance – the *zeitgeist* again.) The resemblance to standard two-choice VI VI matching was interesting, given the many procedural differences between the two. My conclusion: feedback rules. The power-law version of matching is an inevitable result of the feedback implicit in variable-interval schedules.

The most annoying thing about this paper was the behavior of the very Skinnerian journal I sent it to – *The Journal of the Experimental Analysis of Behavior* – which was home to almost all single-animal learning research in those days. The editors were happy with the data, but insisted I omit a short section in the Discussion in which I showed that the pattern of behavior pigeons showed on fixed-interval schedules seems to be optimal. The pigeon's time sense is variable; he cannot judge exactly the time when his next peck will be effective. If he doesn't want to delay reward unnecessarily, he must begin to peck in each interval a little early. He will therefore make some unnecessary pecks. The cost of these must be balanced again the much more severe cost of waiting too long to make that first peck, thus delaying the next reward, and all subsequent rewards. It was all common sense. But the editor objected to the analysis not because it was wrong but because it was somehow inappropriate in *JEAB* – even though all of economics analyses behavior this way. It's not the only way and there are serious problems with relying on a functional analysis (as this is called), but to rule it out meant that what would in fact have been the first such analysis applied to an operant conditioning problem never saw the light of day.

Other experiments tested various implications of the theory of control by time – *temporal control* (TC) – that I had developed beginning with the frustration-effect experiments in Toronto. My idea – not rocket science, by any means – was that temporal control is an aspect of memory. The animal must remember the time marker. If he confuses the most recent time marker with earlier ones, you get the omission effect: the animal (on a fixed-interval schedule) will respond too soon. To judge the time of upcoming reward accurately, he must have some record of the most recent time marker.

Proactive interference

So we did a few memory experiments. The niftiest, perhaps, involved a situation which allowed for good temporal control by a brief neutral (i.e.

not-rewarded) stimulus[3]. It worked like this. The pigeons were first trained on a variable-interval 60-s schedule, a standard, well-understood procedure. They responded at a steady rate. Then in the second phase, every now and then a brief stimulus – three vertical lines– would appear on the response key for 5 s. The stimulus signaled that the next reward would be available for the first peck after two minutes, a fixed-interval schedule. Pretty soon all four birds began to pause after the brief stimulus, showing good temporal control. No overshadowing by the reward here, because, on the VI schedule, it had no signal value and produced no pausing.

Then in the third phase, some of the vertical-line brief stimuli were replaced by three horizontal lines which, like the reward, had no temporal significance. (Pigeons have no difficulty discriminating horizontal and vertical lines in standard experiments.) After the brief horizontal-line stimulus, the VI schedule continued as before. The question: Would the pigeons continue to pause after the vertical-line stimulus, whose significance had not changed? Or would the memory of the horizontal-line stimulus (no pausing) somehow interfere and prevent pausing after the vertical stimulus?

The answer was unequivocal. The pigeons ceased to pause after either line stimulus, indicating some kind of proactive memory interference between them. The birds could, of course have paused after both, which one or two did at first. But that meant delaying many rewards. Or, to put it less teleologically, that meant getting rewarded on many occasions for the very first post-stimulus peck, which led the pigeons to respond sooner and sooner after the stimuli, abolishing the pause. Either way, their failure to treat the two line stimuli differently proved memory interference between them.

The memory idea suggests that a time marker that lasts for a long time should be better than one that is brief. We did a couple of experiments to test this idea. One looked at different feeder durations. Will pigeons trained on fixed-interval wait longer after longer feeder durations? The answer was "yes": the pause was longer, and the "running" response rate higher, after longer feeder durations.

Was it the feeder duration, or just the fact that longer feeding time is a bigger reward? Probably both, because we found in another experiment that longer timeouts produced longer pauses for pigeons and, initially, for rats as well. After a few experimental sessions, the rats pause more or less equally after all durations, from 2– 32 s.

---

[3] Staddon, J. E. R. (1975). Limitations on temporal control: Generalization and the effects of context. *British Journal of Psychology, 66*, 229-246.

The differential pausing does depend on the *intercalation* of different durations. They must be jumbled up together in a sequence, not presented in blocks. If each reward duration is in force for many sessions, pigeons pause similarly after all. Again, the differential pausing indicates memory interference, which is reduced when each duration occurs in a block of trials.

The memory idea suggests that the omission effect should be less, or even zero, if the time delay (FI) is short enough. True: if the time between the neutral stimulus and food is 30 s or less, pigeons' post-stimulus pause is the same as the post-food pause. Every experimental test we could think of confirmed the idea that temporal control depends on memory.

Janice Frank's Toronto PhD research was on discrimination-reversal learning, but it also gave us a chance to look at memory effects. She trained her pigeons on a multiple schedule – two stimuli, alternating every 60 s. Each day, one of the stimuli was paid off on a variable-interval schedule. There was no reward in the other stimulus, and a peck on it delayed the change to the positive stimulus by 30 s. For most of the experiment, the positive stimulus alternated from day to day: Red-plus on Monday, Green-plus on Tuesday, Red-plus on Wednesday and so on. Once they had learned the task, the birds did very well: 90% correct or better each day, usually. The birds learned to 'flip flop'. That is to say, after they had learned if, as a test, we quit rewarding them, they would slowly quit responding to whichever stimulus they decided was positive that day – but only to that stimulus; they rarely sampled the other one. Well-trained birds had learned to pick just one color each day.

More interesting are the memory effects. For example, suppose, after the bird has learned to switch quickly every day (which took a few sessions), we don't run him for a few days. How well might he do on the first day back? On the second day? Well in several tries Janice found that the bird usually did well, maybe even better than usual, on the first day back. But then he did worse on the first reversal on the next day.

Another example: suppose we continue the daily reversals but in the middle of the series switch to a new pair of stimuli, from Red-Green to Square-Circle, say. What should happen now? He has apparently 'learned' to reverse every day, so perhaps the switch will not faze him and he'll continue as before? Well, in fact he does fine on the first day, apparently confirming this suggestion. But he does badly on the second day (first reversal) and on succeeding reversals. Contrary to the hypothesis, he seems to need to relearn the whole thing.

What's going on? Why did the pigeon do well on day one of the new stimulus pair in the last experiment? Why was he better on the first resumption after days of, and why worse on reversals on days after that? All these effects are due to memory interference, what's called *proactive interference. Proactive interference* is the effect of stimulus A on recall of succeeding stimulus B. A familiar example is alternating serve in tennis. At first no one is muddled about who should serve next. But as the game proceeds, the players begin to be uncertain. So also with reversal learning, with the difference that confusion over the 'hot' stimulus (S+) yesterday *aids* performance on the reversal today. Anything which makes it easier to remember what happened yesterday will make it *harder* to *reverse* today. Thus resumption after 'days off' isolates that first day of resumption from earlier reversals, so next day the bird still chooses yesterday's stimulus – and does poorly on the reversal. Similarly, introducing a new stimulus pair means that on the first reversal, yesterday's S+ will be remembered, and interfere with reversal performance.

After very much experience (400+ daily sessions) the days-off effect disappeared, presumably because the now-substantial past history caused persistent memory interference. A manifestation of Jost's Law, the fact that old memories gain strength relative to newer ones with lapse of time? Possibly: obviously, more work is needed to define the numerical limits – how much training? With what? For how long? Nevertheless. this was a beautiful series of experiments that showed subtle and replicable memory effects in individual animals.

The great British biologist J. Z. Young, Professor of Anatomy at UCL, author of *Life of Vertebrates* and *Life of Mammals,* and student of the giant axon of the squid – and now retired – visited Duke regularly during this time. He was doing learning experiments with *Octopus* at the Duke Marine Lab in Beaufort, NC. I can remember having lunch with him at Duke and discussing Janice's reversal experiments – similar procedures have been used to assess the intelligence of *Octopus* and other invertebrates. I was young enough to be much impressed by his intelligence at what I took to be an advanced age. He must have been in his early seventies!

Janice wrote up her experiments and returned to Toronto for her PhD oral exam. All did not go well. No longer a member of the Toronto department, I was not permitted to serve on the examining committee. Were the faculty grumpy about my departure? But why take it out on Janice? I don't know. A Toronto faculty member was chair. Janice failed her oral exam, a shocking – and rare – result. Very few PhD candidates fall at the final fence. No problem with the experiments, surely – we published them in a well-respected journal[4] a few years later. What *was* the problem? Well, for one thing, Janice was profoundly deaf – had been from a very early age. Her speech was hard to understand. She lip-read

---

[4] Staddon, J. E. R., & Frank, J. (1974). Mechanisms of discrimination reversal. *Animal Behaviour, 22,* 802-828.

well, but was always reluctant to ask people to repeat. I must conclude that the committee made insufficient allowance for her disability, since she was certainly smart enough to pass any such examination. I daresay this kind of thoughtlessness is less likely nowadays – one of the few good effects of a victim culture that is often aggressively judgmental (just try sniggering at a transgender person!).

Janice continued to work with me as a research assistant for a few years, but then she got a computer job for which her disability was no disadvantage.

Animal timing became something of a battleground in the late 1970s. On one side were Skinner-trained empirical types – "Just the facts ma'am!" – who were interested in exploring the limits of temporal discrimination. How effective were different kinds of time marker? How quickly can pigeons adapt to changing time-to-reward? How important are procedural details to timing performance? And, how is timing involved in schedule performance in general?

On the other side were cognitive types who were interested not in exploration but in confirmation. Inspired by the 'cognitive revolution' that began in the 1960s, experiments were seen as test-beds for theory. The focus was on a few standard procedures from which exact numerical data might be extracted. One of these is the *peak procedure* which is a variant of fixed interval. Its attraction is that it produces a pattern that fits easily into psychophysical thinking. The procedure works like this. A fixed-interval schedule pays off the first response after I s, where I will be in the range 15 s to perhaps 60 minutes. In the peak procedure, on a certain percentage of trials, there is no reward and the trial extends to perhaps 3 or 4 times I, to be followed by an Inter Trial Interval (ITI) of 15 s or more. On these *extinction trials* the pigeon (or rat) soon learns to quit responding after post-ITI time I. The result is a nice bell-shaped distribution of responding, peaking at I.

But, carpers note, the shape of the distribution must surely depend on the percentage of extinction trials and their length? And, what about the omission effect? Trials begin not with reward, as on fixed-interval, but with a relatively long and often variable Inter Trial Interval. Why variable? Is behavior the same after short ITIs as after long? Possibly "yes" if the ITI is longer than a few seconds? But possibly "no", which just means added noise. Even after working for many years in this area, I'm still not sure why the procedural details are as they are. But again, as with matching and the double runway, the procedure has been tweaked to the point that it yields data conformable to a certain view of the world. No need to meddle further.

Time left, scalar timing and the cognitive pigeon

A more complicated procedure, also designed to fit the cognitive worldview, involves choice. In the *time-left* procedure, after an inter trial interval and lapse of time T, the pigeon must choose left or right. If left, the right key goes out and food reward is available for a peck after $C-T$ s, where C is a fixed value and T is the time since trial onset when the pigeon made his choice. If right, the left key goes out and food is available after a fixed time S.

The aim of time-left is to see if the pigeon can switch his choice at the right time, switching from right (S s to food) to left ($C-T$ s to food) when $C-T < S$. If the bird behaves optimally, then (the assumption is) he must encode time linearly, not nonlinearly as (apparently) implied by Weber's law.

Let's just look at a simple case, where $S = 30$ s and $C = 60$ s. The bird is allowed to choose after a series of times T, from 5 to 50 s. Obviously when the time of choice, T, is 5 s, he's better off choosing S, since 30s is much less than 55 s, the time remaining on the C side. Conversely, when $T = 50$, he's obviously better off choosing C, which only has 10 s to go. It seems pretty obvious that the bird should be indifferent between S and C when $T = 30$. Since the animal is supposed to have 'learned' all the relevant times-to-reward, he just needs to compare them and always choose the shorter. So, in the example, he should be indifferent when $T-C = 30$. Which is what happened in the initial experiments. Score one for the cognitive approach.

But subsequent experiments began to show a bias towards the time-left choice. In other words, the pigeons would often prefer left even when time-to-food was longer than the fixed delay on the right. It soon became clear that the time-left procedure is a conceptual muddle. Cognitive theorists confused the internal scale for representing time with the output scale that directs choice. The cognitive account (bear with me!) goes like this. If time is represented (encoded) logarithmically, as Weber's Law implies, then the first 30 s of a 60-s time to food represent a larger (internal) change than the last 30 s because a log scale is negatively accelerated – steep at first, then flattening out. So S (= 30 s) on the right must be judged as longer than the second 30 s on the left.

But this by no means follows. As I explained earlier in connection with Stevens' power law, people can make judgments according to a power law, even though the stimulus encoding is logarithmic. Indeed, if the exponent of the power law is one, the response will be linear even though encoding is logarithmic. Or, to take a more everyday example, consider the judgment of length. The image of a yardstick on your retina is foreshortened. Its encoded value is smaller the farther away it is. Yet you don't judge a yardstick 10 feet away as smaller than the same one just 1 foot away. The response may be veridical, may correspond to the reality. But the encoding can be quite different.

The time-left experiment is tricky in another way. Any timing model, be it trace or cognitive, can be applied to it in several ways. A trace view goes like this: A trace is left by the inter trial interval. The trace itself is 'tuned' by the animal's history, decaying more slowly after a history when rewards are long delayed than when they occur after a short delay[5]. What must also be learned is the trace value associated with the valued event (food, the reinforcer). If the time-left key (stimulus *C*) is chosen, the time-to-reward is 60-s; so the trace value at that point is when the animal 'expects' the food. But what about the 'standard' (*S*) side? In the usual T-L procedure, only the time-left key stimulus stays the same before and after a choice. The standard-key stimulus changes when it is chosen. What is the effect of stimulus *S*? Does the ITI trace persist even as the stimulus changes? Is the *S* stimulus a 'conditioned reinforcer' so that has a certain value, related to the rate of payoff in its presence? But conditioned reinforcement has problems; a simple reinforcement-rate-equals-value account doesn't work. Or can we ignore the stimulus change and just look at post-ITI time to food? – which time then depends on when the animal chooses the *S* side. It can be from 30s (if he chooses right at trial onset, to 90s, if he chooses at the end of the trial. He should therefore associate (weakly) several trace values with food. It's not at all clear which side the pigeon should prefer.

Never mind models; there is a basic problem with the method. The T-L procedure is asymmetrical in this important way: pecking on the standard side changes the stimulus on the response key; but choosing the time-left side does not. This means that food is delivered in the presence of the time-left choice stimulus but not in the presence of the choice stimulus that produces the standard. Suppose we modify the procedure so that the stimulus changes when either key is pecked. The cognitive argument should not be affected because the time relations are unchanged. But now the pigeons prefer the time-left side at almost all T values. This is also true of the two other possibilities: stimulus change on the T-L side not on the *S* side and no stimulus change on either side. My colleague Dan Cerutti and I played all these variations in a rather elaborate set of experiments a few years later[6].

---

[5] Staddon, J. E. R., & Higa, J. J. & Chelaru, I. M. (1999) Time, trace, memory. *Journal of the Experimental Analysis of Behavior, 71*, 293-301.

[6] Cerutti, D. T., & Staddon, J. E. R. (2004) Immediacy vs anticipated delay in the time-left experiment: A test of the cognitive hypothesis. *Journal of Experimental Psychology: Animal Behavior Processes. 30*, 45–57.

Dan was a wonderful collaborator who worked with me for ten years or so. He eventually moved on to a faculty position in the University of California system. Tragically Dan died suddenly only a couple of years later, leaving a desolated wife and young son, a huge blow.

The time-left procedure is complicated. It is misleading to think of it in cognitive fashion, because that isn't how pigeons work. They don't sit and compare times 'stored in memory' make a subtraction and then 'judge' the shortest time to food. What seems to dominate is reward *immediacy,* the peck-food time, otherwise known as contiguity. The stimulus associated with the shortest peck-food time is usually favored. In the standard procedure, that, of course, is the T-L side, the only one where food occurs in the presence of the same stimulus the animal is choosing.

How does the animal detect immediacy? We don't know all the details. But we do know that some kind of trace is probably involved in all timing. Direct neurophysiological evidence for a memory trace, as for every other suggested timing process, is lacking. But recently, striking circumstantial evidence for a trace emerged in the form of Jazz, an Hungarian Vizsla dog owned by a Scottish couple[7]. Jazz seems to tell time. Every day at about 4:30 he jumps on to the sofa and looks out for Johnny, his master, who always comes home at 5:00. He seems to have a clock that tells him when the boss will arrive.

How does he do it? Perhaps Jazz doesn't have a clock at all? Perhaps he's using *smell* to estimate when Johnny is due to arrive each day. Johnny after all leaves an odor about the house – undetectable to us but certainly not to Jazz whose sense of smell is hundreds of times more acute than ours. And Jazz is obviously a big fan of Johnny's, so his odor is really important. The longer Johnny is out of the house, the weaker his smell. Maybe a certain trace level is the cue that tells Jazz when Johnny is due home?

Wife Christine did a simple test of this idea. One afternoon on her way home an hour or so before Johnny's arrival, she went to his soccer club and got a bunch of his just-used T-shirts. She then went home and waved them around the living room, to "reset" the timing scent. Now the room should smell as if Johnny had just left, not as if he is due to arrive. Amazingly, that day, Jazz showed no anticipation and every evidence of surprise when Johnny arrived at the usual time. Apparently Jazz *is* using the trace level of a scent to estimate Johnny's arrival time.

This single instance by itself does not prove the existence of traces in general. But it does show that a trace can work as a timer. Quite possibly

---

[7] Dogs and Super Senses, NOVA, April 16, 2014.
http://www.youtube.com/watch?v=Ftr9yY-YuYU

animals will use any time-varying signal, external or internal, that is linked to a time marker, as a cue to the passage of time.

In Memorial Hall in the 1960s, laws were the thing: matching, Weber-Fechner and the power law. The cognitive timing types shared this enthusiasm and three decades later rebranded Weber-Fechner timing as *scalar timing*, which shows *time-scale invariance. Scalar* just means that everything is proportional: double the time interval the animal must wait and he waits twice as long with twice the standard deviation (a measure of variability). *Time-scale invariance* is a generalization of the scalar property to, for example, the rate at which a task is learned which depends, it is claimed, only on the ratio of (for example) inter trial interval to trial duration not on their absolute values (the details are complex). Seemingly self-evident explanations were offered for data from the peak procedure and time-left. Timing in fixed-interval schedules is simple: "*An obvious explanation is that the pigeons learned the duration of the interval* between a delivered reinforcement and the next arming of the key and did not begin to peck until they judged that the opportunity to obtain another reinforcement was more or less imminent. [my italics]"

Statements like this raise the hackles of any behaviorist (is that their purpose?). They annoy for two reasons. First, they represent an easy assumption that folk-psychology, the psychology of everyday discourse between human beings, is adequate as a causal account of the behavior of pigeons. The pigeon "judges" – really? Do people even "judge" when they perform simple discriminations? Do you "judge" left vs right direction when you turn left on your regular route home every day? More important, the statement begs the question: the pigeon "learns the duration". Er yeah.... but how? And can we really assume that the bird has a sort of register in which he stores a number representing each time-interval he has learned. Well yes, say many cognitive theorists, invoking the standard human-cognition apparatus of boxes for working and episodic memory.

So how *does* the animal measure these times? Well here also common experience came to the rescue: there is an *internal clock* with a pacemaker (tick). Time is just the count of clock ticks – a model originally proposed for human timing by Michel Treisman at Oxford in 1963. To account for things like the omission effect, the clock can also be partially or completely 'reset', although the causal details for the reset process were not spelled out.

The clock idea also ran into problems. Surprisingly, perhaps, the simplest form of 'noisy' pacemaker clock does not yield Weber's law. To get it, an additional assumption is needed. Physiological evidence for a pacemaker, which, being periodic, should be easy to detect, failed to materialize. Eventually the pacemaker idea was dropped, but no substitute was offered. The trace alternative was simply dismissed. The animals just tell time, apparently. So (all seemed to agree) let's just look at distributions and their properties: did the mean of the peak-interval timing distribution match the training interval? Was its standard deviation proportional to its mean (i.e. Weber-law compliant)? And what does this tell us about the process of learning and the role of timing in it? Who cares? was the implicit response. By seeking procedures that yielded ever-prettier latency distributions to confirm grand Newtonian laws, the cognitive approach diverted attention away from the larger question of how animals actually learn and adapt to temporal reinforcement schedules.

Ideas in psychology have a Darwinian 'fitness' that depends only partly on their experimental truth. 'Frustration' is a striking example. Amsel's original paper has been cited over 400 times and his follow-up review, nearly 1400. The various refutations of 'frustration theory' have never been cited more than 50 times. One reason for the disparity is that the number of neo-Hullian researchers, i.e. Amsel's co-religionists, is larger than the comparable Skinner group – never mind the small subset of operant researchers with theoretical interests. I'll suggest why in a moment. And of course, conclusive disproof will rarely be cited simply because it *is* conclusive. A piece of research that removes the foundation from under a whole area of research, to the extent that it is truly effective will be heard from no more, simply because its subject has ceased to be of general interest.

But the main reason that 'frustration theory' persists is surely because it is so familiar. An influential textbook in 1975 misunderstood the reason for the pause on fixed-interval schedules in our 'frustration' experiments, and also assumed that any effect we showed must be in addition to, or 'masking', a frustration effect – an effect whose existence was apparently beyond doubt. A few people are still studying the FE, even though it should be abundantly clear by now that if the 'frustration effect' exists at all it is much smaller than the disinhibitory effect we proposed as an alternative. Same for the pacemaker clock. It will not go extinct, simply because it is so intuitively appealing.

The 'softer' sciences are inescapably prone to this balancing act: disproof must be absolutely overwhelming before a false but plausible idea can be fully interred. Otherwise fallacy prospers. But conclusive disproof is often difficult to impossible outside the hard sciences. Hence what might be called 'the immortality of the plausible'.

There is an analogy to what is called in genetics *recurrent mutation*. A given gene may occur in many forms (alleles). Some mutant alleles occur rarely, but some recur at a substantial rate. The polymath UCL biologist J. B. S. Haldane many years ago looked at the balance between mutation rate and selection rate. He showed that even if selection acts against them, a certain fraction of the population will always possess these recurrent alleles. So also with plausible ideas in soft science. Selection – experimental data – acts against them, but plausibility ensures that although they may be weakened, many will continue to believe them. Richard Dawkins in his bestseller *The Selfish Gene* noted the similarity between genes and ideas, which he called memes. But as far as I know no one has pointed out the analogy to false but attractive ideas: recurrent memes.

J. B. S Haldane was a fascinating character. Born into a patrician family, he was a brilliant biologist even though he had no formal qualifications – his Oxford degree was in math and classics. Politically he once claimed to be an anti-government Jeffersonian. In other respects he was in the same left-wing tradition as Bernal and Hogben. A Marxist and admirer of Lenin and even Stalin, when he retired he left Britain and became a naturalized Indian, partly because of the Suez crisis and partly because of a wish to "never again wear socks". He strongly disapproved of the British government's attempt (feeble and ineffective though it was, largely owing to the opposition of the U.S.) to recapture the Suez Canal from Egyptian strongman Gamal Abdel Nasser. Haldane spent the rest of his life in India, where he was much loved. But during my first year at UCL he was regularly visible, repairing most afternoons to the Marlborough Arms to chat with students.

There is one other reason that fallacies persist in the softer sciences. The big difference between Skinner- and Hull-inspired learning research is the experimental method: individual animals versus comparisons between groups. One might think that group research being harder, or at least more labor intensive, should have fewer practitioners. But the opposite seems to be the case. The reason for its popularity may be not the method itself, but how it fits the incentive structure of contemporary science. Science used to be a vocation; it still is for a few. But it is also for most scientists a *career*. And a career means salary and promotion. Both now depend on *productivity*, which is measured by the scientist's ability to garner research funds and by publication in prestigious journals. No grant, few publications, equals no tenure and no salary. The between-group method can help.

Real science often means years of failure. Just check out biographies of great 19th century scientists like Darwin or Humphry Davy, who labored for years on some projects before making their discoveries. Darwin worked for eight years on barnacles before publishing and some 23 years on his great theory – and even then only published what he thought was still an incomplete account because it looked as if Alfred Russel Wallace might scoop him. Davy took many years and dozens of experiments before he nailed down the new elements sodium and potassium. A career scientist now must show progress over a two- or three-year span. What he needs is a method that guarantees a publishable product. Real science can promise no such thing. Best find a method that can.

So what is the reason for the popularity of the between-group method? Between-group studies use a statistical method called Null Hypothesis Statistical Testing (NHST for short). In its simplest form, the NHST method requires an experimental group, which gets whatever treatment is being studied, and a control group identical in every way to the experimental save for the treatment. Data – means and variances (variability) – are compared, and standard statistical techniques applied. If you're lucky there is a difference between the groups that is 'statistically significant' at the 5% level' or better. All this means is that if the two groups are really the same, i.e. no effect of the treatment, the difference you found would have occurred by chance only 5% of the time. But this of course, *obviously*, means that out of twenty experiments to test a false hypothesis, i.e. experiments where there really is no effect, you are likely to find a significant effect in one. As it has been practiced for the past 80 years or so, the NHST method in effect guarantees a (statistically) 'significant' result with probability of one in twenty even if every hypothesis is wrong. Every experiment should fail, yet the method guarantees that 5% may seem to succeed.

The method fulfils a social need if not a scientific one. By using this method, a diligent researcher can be sure of a modicum of success – in the form of publishable, because statistically significant, research – even if every hypothesis is wrong and every study really a failure. Of course, it means that very many experiments in the social science literature that have used these methods likely come to false conclusions and will not give the same results if someone tries to repeat them. And there will be more of these false results the greater the number of scientists using the NHST method.

This is a huge flaw in the method. But few paid much attention until it began to impact the bottom line: in drug studies. Drug companies look to the scientific literature for drugs they can manufacture and market. But "When Bayer tried to replicate results of 67 studies published in academic journals, nearly two thirds failed"[8] wrote the *Wall Street Journal* in 2011. No matter how predictable this should have been, simply from the flawed

---

[8] Gautam Naik: Scientists' Elusive Goal: Reproducing Study Results *Wall Street Journal*, 12/02/11. See also Chapter 9 in *The New Behaviorism* (2014).

statistical methods used in these drug trials, this result was a shock. Attention is now being paid to all the artifactual results that have been perpetrated. 'Replication' is the new buzzword. But whether logic can overcome the unfortunate social structure of science that has evolved since WW2 remains to be seen.

These reservations are less relevant to engineering and applied science – science where the objective is set in advance and the methods to be used are for the most part familiar. Curiosity-based – basic – research, however, has been much corrupted by the career model.

Timescale Variants

Timing was a contentious area in the 1990s. Here is a slightly edited version of an unpublished critique I submitted (twice!) to the Psychological Review. *Two issues were: Is* timescale invariance *a new psychological principle, or is it mostly the old Weber-Fechner law with a new name? Is there, as claimed, a sharp transition between not-responding and responding when animals adapt to a periodic schedule or do they just slowly accelerate their rate of responding?*

Gallistel & Gibbon (2000) propose an ambitious synthesis of experimental research on conditioning and interval timing. They claim that interval timing by animals is relative so that changing the time scale by a constant factor leaves performance unchanged. But three sets of experimental data – parametric studies of fixed-interval behavior, studies of greater-less discrimination of time intervals and the fixed-interval "scallop" itself – all show that interval timing does depend to some extent on absolute duration. Their 'new' principle of *timescale invariance* (TSI) is Weber's Law in a new dress, and like Weber's Law is only approximately true.

TSI is the heart of their new synthesis: "A ... feature of these models is their timescale invariance, which the authors argue is a very important property evident in the available experimental data....A conditioning result is timescale-invariant if the graph of the result looks the same when the experiment is repeated at different timescale, by changing all the temporal intervals in the protocol by a common scaling factor...data are timescale-invariant if the normalized data plots are superimposable." G & G apply this idea to two kinds of data: interval timing and associative conditioning.

Three kinds of timing data show that TSI is at best an approximation. Perhaps the simplest is the steady-state pattern of responding (key-pecking or lever-pressing) observed on fixed-interval (FI) reinforcement schedules, that is, a schedule in which the first post-reinforcement response T s after reinforcement produces reinforcement. Citing Schneider (1969) G & G write "When responding on such a schedule, animals pause after each reinforcement and then resume responding after some interval has elapsed. *It was generally supposed that the animals' rate of responding accelerated throughout the remainder of the interval leading up to reinforcement.* In fact, however, conditioned responding in this paradigm...*is a two-state variable* (slow, sporadic pecking vs rapid, steady pecking), with one transition per inter-reinforcement interval..." (p. 293, my italics)

This conclusion does not follow from Schneider's result. Reacting to reports of "break-and-run" FI performance under some conditions,

Schneider sought to show it in a more elegant, or at least more statistical, way than the simple inspection of cumulative records. He found a way to identify the point of maximum acceleration in the FI "scallop" by using an iterative technique analogous to attaching an elastic band to the beginning of an interval and the end point of the cumulative record, then pushing a pin, representing the "break point", against the middle of the band until the two straight-line segments best fit the cumulative record. The post-reinforcement time (*x*-coordinate) of the pin then gives the break point for that interval. He showed that the break point is roughly .67 of interval duration, with standard deviation proportional to the mean (the Weber-law property).

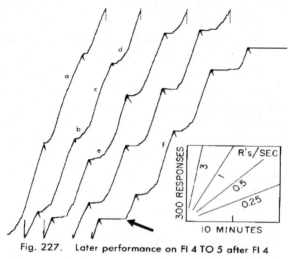

Fig. 227.   Later performance on FI 4 TO 5 after FI 4

Figure 10.1 Cumulative records for a single pigeon long-trained on fixed-interval 4 min. Each peck (there are thousands) pushes the record up a step; time moves it from left to right. The hash marks show food delivery followed by a 5-min timeout (the recorder is stopped during this time). The usual pattern is wait (flat line) followed by accelerating responding – the "FI scallop". Only one interval shows a sharp transition between waiting and responding (arrow).

But this finding is by no means the same as G & G's assertion that the FI scallop *is* "a two-state variable." Schneider showed that a 2-state model is an adequate approximation; he did not show that it is the best or truest approximation. A 3- or 4-line approximation might well have fit significantly better than the 2-line version. To show that the process *is* 2-state, Schneider would have had to show that adding additional segments produced negligibly better fit to the data, and that the goodness-of-fit was the same at long and short FI values.

Nevertheless, G & G claim that the FI scallop is always an artifact of averaging: "As a result [of variability], averaging over many interreinforcement intervals results in the smooth increase in average rate of responding..." True, but irrelevant: the raw, statistically unmodified, data show that the FI scallop plainly is *not* an averaging artifact. A glance at the many non-averaged individual FI cumulative records in Ferster and Skinner (1957) shows clear curvature, particularly at longer FI values (> 2 min or so). For example, Figure 10.1 (F & S Fig. 227) shows that of the 16 or so intervals illustrated, only one record (arrow) looks at all like break-and-run. Most FI cumulative records show a clear scallop, although some, particularly at shorter FI values or after protracted training, do not. G & G's "it was generally supposed" comment above is in the spirit of "Who are you going to believe, me or your lyin' eyes?"

Moreover, even Schneider's data show the effect of absolute duration. In Schneider's Figure 3, for example, the time to shift from low to high rate is clearly longer at longer intervals (256, 512 s) than shorter ones. On FI schedules, apparently, absolute duration *does* affect the pattern of responding. If there is a difference in the shape of cumulative records at long and short FI values – and there is – timescale invariance is violated.

There are two other more direct tests of timescale invariance on temporal schedules. Using the popular peak-interval procedure, Zeiler & Powell (1994) looked explicitly at the effect of interval duration on various measures of interval timing. They conclude "Quantitative properties of temporal control depended on whether the aspect of behavior considered was initial pause duration, the point of maximum acceleration in responding [break point], the point of maximum deceleration, the point at which responding stopped, or several different statistical derivations of a point of maximum responding...Existing theory does not explain why Weber's law [the scalar property] so rarely fit the results..." (p. 1; see also Lowe, Harzem, & Spencer, 1979, and Wearden, 1985, for other exceptions to proportionality between temporal measures of behavior and interval duration.) Like Schneider, Zeiler and Powell found that the break point measure was proportional to interval duration, with scalar variance (constant coefficient of variation), and thus consistent with timescale invariance. But no other measure fit the rule.

Moreover, the fit of breakpoint is problematic since it is not a direct measure but is itself the result of a statistical procedure. It is possible, therefore, that the fit of breakpoint to timescale invariance owes as much to the statistical method used to arrive at it as to the intrinsic properties of temporal control. Even if this caveat turns out to be false, the fact that every other measure studied by Zeiler and Powell failed to conform to time scale invariance surely rules it out as a general principle of interval timing.

Perhaps the most direct test of the timescale invariance idea is an extensive series of experiments on time discrimination carried out by Dreyfus, Fetterman, Smith & Stubbs (1988) and Stubbs, Dreyfus, Fetterman, Boynton, Locklin and Smith (1994). The usual procedure in these experiments was for pigeons to peck a center response key to produce a red light of one duration that is followed immediately by a green light of another duration. When the green center-key light goes off, two yellow side keys light up. The animals are reinforced with food for pecking the left side-key if the red light was longer, the right side-key if the green light was longer.

The experimental question is: How does discrimination accuracy depend on relative and absolute duration of the two stimuli? Timescale invariance would presumably say that accuracy depends only on the ratio of red and green durations: accuracy should be same following the sequence red: 10, green: 20 as the sequence red: 30, green: 60, for example. But it is not. Pigeons are better able to discriminate between the two short durations than the two long, even though their ratio is the same. Dreyfus et al., and Stubbs et al. present a plethora of quantitative data of the same sort, all showing that time discrimination depends on absolute as well as relative duration.

The principle of timescale invariance is true only of a limited set of interval-timing data. It is in effect indistinguishable from Weber's law as it applies to time and, like Weber's law, is on approximately true. It does not qualify as a grand new principle.

The principle fares little better as applied to conditioning. The idea that a conditioning measure such as "reinforcements to acquisition" depends only on the ratio of trial duration to intertrial interval ("ratio invariance") is very broadly true (Gallistel & Gibbon, Figure 9), but Figure 9 is in log-log coordinates that conceal substantial variability in the dependent variable. Moreover, recent studies aimed at a direct test of ratio invariance (e.g. Holland, 2000; Lattal, 1999) show it to be valid only over a limited range of times. These data support only a monotonic relation between rate of conditioning and the ratio of trial duration and intertrial interval, not a quantitative law. And the linkage between TSI as a timing principle and ratio-invariance as a conditioning principle is problematic. For example, Holland's experimental evidence leads him to conclude that "associative strength and temporal information are separately represented" (2000, p. 134). Timescale invariance is not new enough or true enough to deserve a new name.

References

Dreyfus, L. R., Fetterman, J. G., Smith, L. D., & Stubbs, D. A. (1988) Discrimination of temporal relations by pigeons. *Journal of Experimental Psychology: Animal Behavior Process, 14,*349-367.

Ferster, C. B., & Skinner, B. F. *Schedules of reinforcement.* New York: Appleton-Century-Crofts.

Gallistel, C. R., & Gibbon, J. (2000) Time, rate, and conditioning. *Psychological Review, 107*, 289-344.

Holland, P. C. (2000) Trial and intertrial durations in appetitive conditioning in rats. *Animal Learning and Behavior, 28,* 121-135.

Lattal, K. M. (1999) Trial and intertrial durations in Pavlovian conditioning. *Journal of Experimental Psychology: Animal Behavior Processes, 25*, 433-450.

Lowe, C. F., Harzem, P., & Spencer, P. T. (1979) Temporal control of behavior and the power law. *Journal of the Experimental Analysis of Behavior, 31*, 333-343.

Schneider, B. A. (1969) A two-state analysis of fixed-interval responding in pigeons. *Journal of the Experimental Analysis of Behavior, 12*, 677-687.

Stubbs, D. A., Dreyfus, L. R., Fetterman, J. G. Boynton, D. M., Locklin, N., & Smith, L. D. (1994) Duration comparison: relative stimulus differences, stimulus age and stimulus predictiveness. *Journal of the Experimental Analysis of Behavior, 62,* 15-32.

Wearden, J. H. (1985) The power law and Weber's law in fixed-interval post-reinforcement pausing. *Quarterly Journal of Experimental Psychology, 37(3)*, 191-211.

Zeiler, M. D., & Powell, D. G. (1994) Temporal control in fixed-interval schedules. *Journal of the Experimental Analysis of Behavior, 61*, 1-9.

Chapter 11: Bird song, evolution and analogy

My parents stayed on in Rhodesia, now independent Zambia, after 1964. In 1969 they decided to emigrate to the U.S., to join us in Durham. They shipped four large black-painted crates of their belongings, plus a smaller box made of an African wood (my father called the wood *mukwa*, a common Bemba word – but I think it means any wood, not necessarily the nice hardwood he used for the box). But only the four painted crates made it to Durham, NC, where I had been teaching for two years. The missing box contained most of their personal stuff – valuables, photographs and so on. It was stolen in the New York docks. A year or so later we got a strange communication that promised to restore the lost box, but it was pretty obviously some kind of scam. So, not a good beginning.

They rented a pleasant condo near the Lakewood shopping center close to us and my father soon got a job with the local Coman Lumber Company, working in their stores. I acquired many nice scraps of wood from there, and made a large bookcase, a big box for Woozle's toys and a loudspeaker cabinet. My father bought a handsome car, a 1966 Ford. All went well for a while but Dad found the physical work at Coman hard. I found out much later that DB was not perhaps as welcoming as she might have been. And I was always busy. My father, particularly, missed his social life, I think. So within a couple of years they returned to England. I inherited the Ford, but it developed a tendency to overheat and soon had to go.

On the lab front, technology was beginning to change at an accelerating rate. When I arrived at Duke in 1967 I briefly inherited a LINC (named after its origin in the MIT Lincoln Lab – also for the acronym Laboratory Instrument Computer) although I never used it. LINC was a project supported by the National Institute of Health. It was a small computer – about the size of a dishwasher – designed to aid electrophysiological research. My predecessor Al Boneau had got it through a grant. It was soon returned to him – he had moved to NC State, just down the road. The LINC, like so many government-inspired projects, was in fact a bit premature. Real computers were not ready to take over operant labs for a while. We continued to use electromechanical relays, clocks and timers for a few years. In the meantime, a couple of companies began to market logic systems based on individual transistors which did the same job faster and used less space. These had a brief life.

The next stage was the appearance in the late '70s of small single-board computers. Communication was via teletype, basically an electric typewriter that could communicate and print 8-channel punched paper tape. These early computers were mostly based on the Intel 8080 chip but we got a couple of 6502-based KIM devices which had to be programmed directly in machine language. Storage was of course limited. The pace quickened. I remember something called the Northstar Horizon, which was in fact a kit based on the Z80 chip. It had couple of floppy disk drives (5 ¼ in – smaller than the previous 7-in standard!) and a hard disk with 5 megabytes (!) of storage. It may have had as much as 64K of RAM. Young Nick, beginning a lifelong involvement with computers, helped the graduate students put together one of these. It ran the CP/M operating system, which was primitive but pretty good. We bought the hardware literally from a house in a field near Greensboro. An enterprising overalled entrepreneur had bought a bunch of Northstars and sold them from his farmhouse, complete with pot-belly stove and Grandma doing the accounts.

The story of CP/M is an interesting commentary on what it takes – besides a great dollop of luck – to succeed in American tech business. It was the best operating system around for a while, invented and sold by Gary Kildall's company Digital Research. Kildall became quite rich quite quickly – and perhaps less hungry. He had a chance to sell CP/M to IBM for their new personal computer, the one that jump-started the PC industry. Kildall was not a deal maker; apparently he left much of that to his wife. In any event, through lack of interest plus circumstances and the force of the competition, Kildall lost to Bill Gates, who *was* very hungry. Gates bought another, and imperfect operating system called QDOS, reworked it as PC-DOS and got the IBM contract, with the results we all know.

After the era of minicomputers dawned we went through many of them, the Digital Equipment Corp. PDP series, first PDP 8 then PDP 11. I learned my second computer language, FOCAL, designed to fit the tiny memories of these early PDPs. A SUN Microsystems machine was powerful enough to the run the Psychology Department email for a while. Finally, everything was run on boring IBM PCs. I do remember one of the graduate students having a hard time with the transition from PDPs because he could no longer control by panel switches the bit-by-bit contents of the machine registers. Typing commands from a keyboard felt like loss of control.

Word processing was invented. I first encountered it on a visit to California in the late 1970s. I visited the identical twin brother of a Duke colleague who worked for a biomedical company. (That was a strange experience; they even sounded the same. Luckily the California one had a mustache.) He showed me a desk-sized computer setup made by Wang that ran a word processor. The secretary (they existed then) showed me how to delete and add words, move a paragraph and so on. I was entranced. What a great device! The Wang setup cost several thousand dollars but I found a simple 6502 computer (I think it was a Commodore

PET) that ran a word processor called WordStar. Some of WordStar's odd pre-mouse keyboard commands live on still in my current version of Microsoft WORD.

In the meantime, research in the lab took on a more biological cast. There was the theme of adaptation, both *phylogenetic*, across geological time, and *ontogenetic*, the various kinds of learning. Can the idea of adaptiveness help us understand learning? There were a few one-off studies following up leads that looked interesting.

Intention movements and supernormal stimuli

One lead came from grad student Mike Davis, who noticed something odd about flocks of feeding pigeons. Once in a while, one bird would take flight and rouse the whole flock to follow. But at other times, a bird could take off with no effect on others. Mike looked very carefully to see if there was a difference between what we called "alarm flights" and others. After taking many movies and studying alarm takeoffs in greatest detail, Mike concluded: there are no differences. Alarm flights are not different from non-alarm. The difference is in what happens *before* the flight. A spontaneous (i.e. unprovoked by some alarming stimulus) flight is always preceded by *intention movements* – wing lifts, movements of the head and so on. These serve to let the flock know the bird is about to fly and no notice is taken[1]. But if the bird takes off without these preliminaries, the flock is alarmed and takes off itself. There is a human analogue. Imagine a crowded lecture room. A student methodically gathers up her books and leaves in the middle of the lecture. No one pays much attention. But if she starts to her feet without warning and leaves, everyone will notice.

The analogy between learning and evolution by natural selection has become more and more compelling over the years. *Variation* – of gene pool or behavioral repertoire – is what allows *selection* (reinforcement) to produce adaptation (operant learning). If the behavior you want never occurs, it can't be selected. It is no surprise, therefore, to find that many learning phenomena have parallels in evolution. One that struck me was what is called peak shift in stimulus generalization. As Norm Guttman showed, if a pigeon is well trained to peck at a spectral green light (550 nm, to be precise) on an intermittent schedule (variable-interval), he will continue to peck for an hour or two even if he gets no food. If during that hour, the key is successively illuminated with other wavelengths, from 450 to 75 nm, say, he will peck each at a rate that depends on the difference between the new wavelength and the training wavelength. The

---

[1] Davis, J. Michael. Socially-induced flight reactions in pigeons. Ph.D. Dissertation (Duke, 1973), published as Davis, J. M. (1975) Socially induced flight reactions in pigeons. *Animal Behavior* 23, 597-601.

result is a *generalization gradient* with a peak at 550 nm, declining symmetrically on either side.

Suppose now the procedure is changed from 60 min of green light to 1-min-intervals alternating between green and, say, white light. The VI schedule continues in the green, but pecks get no food in white. Then the green light is replaced with a range of wavelengths as before. The white component is unchanged. The result? A symmetrical gradient, as before, but more responses per minute in each stimulus (*behavioral contrast*).

All this was familiar in the 1950s. But what if you train the pigeon to discriminate not between green and white but between green and something close to green, say 550 (S+) vs 600 (S-)? H. M. Hanson showed in 1959 that the effect of training the animal to discriminate between two adjacent wavelengths is to shift the resulting gradient away from the negative stimulus. 550 S+ and 600 S- might result in a gradient with a peak at 500. In other words, paradoxically, the training causes the animal to prefer (or at least respond faster to) a stimulus it has never seen than to the training stimulus. This is called *peak shift.*

Ethologists found something similar in the instinctive behavior of many species. Birds like big eggs – no learning involved. In many species, if an artificial, larger-than-normal egg is introduced into the nest, the sitting bird will prefer it over the others. Ground-nesting terns, for example will retrieve a large egg that has rolled out of the nest before getting a normal-sized one. Some male butterflies, attracted by the wing-beat of a female, will prefer a faster-than-normal wing-beat over a slower one. The peak-shift experiments suggested why[2]: *asymmetrical selection*, that is, an evolutionary environment which, like the conditioning procedure, disfavors stimuli on one side of the norm more than stimuli on the other. In the case of the eggs, smaller-than-normal ones are likely to be non-viable, so retrieving one may well be a waste of time and nest-space. Normal eggs are OK – and abnormally large ones are never seen in nature because the bird is too small to lay them. So, no selection against retrieving them, but selection against retrieving small ones. The result: preference for too-large eggs. I made the same argument for the butterflies. A slow wing-beat might be a bird which wants you for lunch. Avoid that. Very fast wing-beat is rarely encountered, selection shifts the peak to a preference for the faster one.

My little paper on supernormal stimuli led to my first encounter with the hypersensitivity of eminent persons. I got a testy letter from Ernst Mayr, a distinguished professor of evolutionary biology at Harvard. He complained that I had not cited him and gave page references to his 1963

---

[2] Staddon, J. E. R. (1975). A note on the evolutionary significance of supernormal stimuli. *American Naturalist, 109*, 541-545.

book *Animal Species and Evolution*. I checked the pages and could see nothing related to stimulus generalization, a topic about which I would not have expected Mayr to know anything at all. Nor could I find any mention of asymmetrical selection pressure, the main evolutionary implication of my paper. So I wrote back to say that I was not unaware of Mayr's work. Indeed, I knew and had learned much from his 1963 book – which was a sourcebook for me and many others interested in evolution and behavior. But I could not see that he had in any way anticipated what I had to say in my article, so I could not apologize for failing to cite his book. I got no response and I'm still puzzled by Mayr's letter.

I had a similar experience a few years later with another eminence: one Crawford H. Greenewalt. This one was even more incomprehensible than the brush with Mayr. In the mid-1970s, I became interested in bird song as a result of a collaboration with two UNC Chapel Hill comparative psychologists, Andrew King and Meredith West, a couple who had set up a little research station at their home near Saxapahaw, west of Chapel Hill. They were interested in the behavior of birds, especially the Brown Headed Cowbird. I had an interest in psychoacoustics and had a good background from Smitty Stevens' indoctrination in Memorial Hall. DB and I helped them in various ways with their work. We kept birds in isolation (from other birds, not from us) at our home to see how their calls might change over the winter, for example. We found that starlings learned to imitate speech. But the main product of the collaboration was a gadget to analyze bird calls in an era before computers could do it. Computers now are fast enough to do the job with an analog-to-digital converter and suitable software 'app'. The only gadget then available was something called a sound spectrograph, made by the Kay corporation and sold for a hefty price. It produced a printed visual record of frequency, on the vertical axis, versus time, on the horizontal. Most bird song is close to a pure tone, so the spectrograph record is mostly a single line, interrupted by breaks between calls. The sonograph was slow, the record was brief, just a second or two, and appeared only after a delay.

Bird song

An acoustic signal is a wave which oscillates between positive and negative values on the air-pressure axis. Since most birdsong is close to a pure tone, it occurred to me that most of the information is in the times between zero crossings, when the sound pressure flips from positive to negative. Given a steady pure tone, for example, the time between successive positive-going crossings is inversely proportional to the tone frequency. So I proposed a device that would make a dot on an oscilloscope screen whenever the audio signal went from positive to negative. The vertical coordinate of the dot would be directly related to time since the last zero crossing and the horizontal coordinate would be time since the beginning of the recording. In this way, I could get a record with time on the x-axis and inter-zero-crossing time (1/IZC is frequency, of course) on the y-axis. The resulting pattern could be photographed with a Polaroid camera or, preserved on a storage oscilloscope. This arrangement could be much more flexible than the sound spectrograph because the readout is immediate and the time scale could be varied at will.

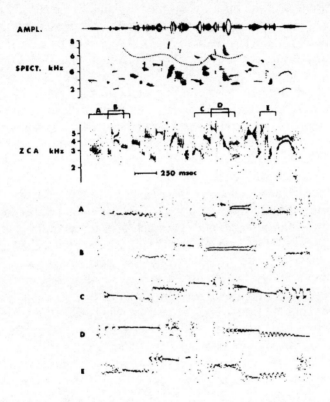

Figure 11.1
2.5-s clip of the song of the Lapland Longspur (*Calcarius lapponicus*), one of the most intricate and beautiful whistled bird songs. Song amplitude (intensity) at the top, sonogram below, ZCA traces below that. A, B, C etc. are expanded segments of the song, showing its rich structure.

It was one thing to think of the idea, quite another to actually make a device that would implement it. The essential ingredient was a biomedical engineering student called Bob Bruce, recommended to me by my friend Myron "Mike" Wolbarsht. Mike was an eccentric and multitalented character with appointments in BME and the Department of Ophthalmology in the Medical School. Bob, despite his undergraduate major in English, was a genius at electronics. Using just two chips, he put together a little circuit board that did just what I needed. Then, with two more collaborators we published a paper showing just what the *zero-crossing analyzer* could do[3].

The ZCA display revealed patterns in many bird songs that are not visible in spectrographic displays. Figure 11.1 shows a rather striking example from a little bird called the Lapland Longspur. The blurry spectrogram conceals much structure which is revealed in the zero-crossing display.

I still have no idea whether this structure is significant or just an artifact of the zero-crossing method. Now, it would be easy to generate test songs with different zero-crossing properties to see if, for example, particular artificial songs might be more effective than the natural song – that is, be supernormal stimuli, showing that the zero crossings are the key information in the signal. I failed to get a grant for more work on this rather tech-heavy topic and turned to other things. More research is definitely needed on this one!

I was sufficiently excited by the properties of the ZCA that I looked into the patent possibilities, even though the potential market for such a device – bird song research and not much else – was minuscule. A few inquiries knocked that one on the head. I also approached the Kay company, makers of the widely used sound spectrograph, to see if they might be interested. Not at all, they said. Nevertheless, within a couple of years, they produced their own zero-crossing device.

We were not the first to look at zero crossings. Crawford Greenewalt literally wrote the book on the acoustics of bird song: *Bird Song. Acoustics and Physiology* published in 1968. Greenewalt was not your run-of-the mill academic researcher, but the head of the mighty DuPont Corporation and a man of wide interests. He cooperated with the legendary Harold Edgerton to make high speed photographs of hummingbirds, for example. He discussed zero-crossings in his book, but had no easy way to record them. He didn't really need 'easy' since he had all the technical resources of DuPont to do what he needed.

We of course cited Greenewalt in our paper, but he nevertheless wrote to me complaining that we had not done so. I have absolutely no idea why. Perhaps it was because the *Zeitschrift* put in the references one after the other, not, as is usually done, with each one beginning a new line, so

---

[3] Staddon, J. E. R., McGeorge, L. W., Bruce, R. A., & Klein, F. F. (1978). A simple method for the rapid analysis of animal sounds. *Zeitschrift für Tierpsychologie, 48*, 306-330.

"Greenewalt" appeared in the middle of a line. Mr. G surely didn't read the paper or else he would have seen the entire second paragraph devoted to his work on zero crossings. His name appears repeatedly in our paper. This one really is a mystery, and another testimony to the reputational jealousy of the eminent.

My dip into bioacoustics continued. Cowbirds are brood parasites like cuckoos, they lay their eggs in the nests of other birds – up to 300 species of other birds, according to some counts. They're really good at it! West and King used the ZCA to distinguish elements in the mating call of the male and show that some combinations were more effective than others. The males, raised by other species, have no one to learn their calls from. They are built-in – instinctive. But they are modifiable. A male kept in isolation will in fact sing a song that is more effective than a male that hangs out with other males. But the other males don't like it and attack the sexy male until he changes his tune to conform. Eastern and Western cowbirds have slightly different songs. Western females like Western songs and Eastern females Eastern songs. But an Eastern male confronted with Western females will modify his song to accommodate their taste. Female cowbirds rule.

My two other co-authors on the ZCA paper were Fritz Klein, a Psychology Department technician and Lee McGeorge a zoology student. In those days, Duke had four technicians. Now we, and most other psychology departments, have just one – the computer guy – as what used to be experimental psychology has either vanished or morphed into neuroscience.

I first heard from Lee Wilson McGeorge via a note. When we finally met I was surprised to see not the male student I had expected but a rather beautiful young woman. Lee is from Memphis, Tennessee – southerners often seem to give girl babies boy names. She was a zoology graduate student who had just come back from Madagascar where she worked on lemurs. She worked there with another student who was, understandably, devastated when she apparently left him on her return to the U.S. She had consulted me on some acoustic issue and helped on the ZCA paper. But she mainly worked at the Duke lemur center, which has become a world-renowned conservation site.

At this time a British nature writer, Gerald Durrell, brother of novelist Lawrence and famous for best-sellers such as *My Family and Other Animals,* was a 'Visitor' to the Center. He met Lee on one of his visits and later contacted her with the suggestion that she take up a scholarship that would allow her to spend time at his Jersey Zoological Park. She accepted, but of course it was a ruse: the donor was Gerald and the only candidate was Lee. Once she got to Jersey, he courted her, sending her notes illustrated with little drawings like the ones in his books and stuck

THIRD EDITION

Psychology of Learning and Behavior

Barry Schwartz

***Left*: A Brown-Headed Cowbird on our deck in Kent Street.**
***Below*: One of JZ Young's octopi.**
***Right*: A pigeon showing 'superstitious' flying-hopping interim behavior, in a demo apparatus we used for classes at Duke.**
(Photo by Bill Boyarsky)

under her door. They were eventually married in 1979 and she became Lee Durrell, soon to become famous for her TV appearances with Gerald in a variety of animal and conservation programs. There was, of course, a certain disparity in age. Gerald died relatively young at 70 in 1995, but Lee is still active in the *Durrell Wildlife Conservation Trust*.

In 1978 I became an American citizen. The ceremony was held in Greensboro and the address afterwards was given by Kenneth Pye, then Chancellor of Duke. I was quizzed, as all applicants are, by a young man from the State department. "You are a professor" he said, smiling, "so I will have to ask you serious questions". He proceeded to ask me obscure questions about the Continental Congress and the Articles of Confederation most of which were beyond me. My sponsor was the all-knowing Mike Wolbarsht who was able to answer most of them. It was all in good fun, of course, and the ceremony as a whole was a moving experience. Especially so for the other new citizens, in all colors and from all nations. The group of a hundred or so happy people was a diversity officers delight.

Afterwards I was accosted by a young man with a microphone from the local public radio station. It became clear that he was looking for bad news, someone to complain about the process. He was obviously disappointed to find only people who were delighted to be new Americans. I was reminded of this experience recently listening to the BBC news on the radio. Apparently a woman jockey had just won the most prestigious race in the Australian calendar, and had complained about discrimination. Unable to get hold of her, the reporter was interviewing an eminent British female jockey. She answered every question in the negative. No, people had been very helpful, especially the men; no, she couldn't say she had felt any prejudice. And so on for several questions, until the reporter reluctantly gave up.

The Duke Lemur Center was founded the year before I arrived at Duke by two Duke faculty, Peter Klopfer, an ethologist in the Zoology Department and John Buettner-Janusch, a biochemical anthropologist. The two got a large grant from the National Science Foundation to study the behavior, evolution and blood chemistry of lemurs. Lemurs are interesting not because they are cute (although that no doubt helps keep the Center reasonably well funded), but because they are primates which have evolved in Madagascar in isolation from the African mainland for some forty million years. The Center has grown into a major conservation resource. I was briefly on its Steering Committee when the head was Elwyn Simons, a primate paleontologist with, as far as I know, no professional experience with live animals. The committee meetings – when he was not at a study site in Egypt – were basically monologues by Elwyn. Nevertheless, with a few ups and downs, the Center survived and even prospered. Peter K and I interacted quite a lot and I taught his animal behavior course when he was away on sabbatical. He lives in rural part of Durham County where he and wife Martha founded the Carolina Friends School, to which both my children went at various times.

I first encountered the Lemur Center's other founder, B-J as he was called, at faculty gatherings. He was one of the brightest people I have known. I was especially impressed by his wit and cleverness in repartee. It was this, I suspect, that caused his eventual downfall. B-J's wife Vina ran his lab and, I think, much of his life. In 1973 he moved from Duke to New York University. A few years later his wife died and it seems that his life rather unraveled after that. B-J was accused by some of his own graduate students of making drugs in his chemistry lab. The charge has always seemed incredible to me. He didn't need the money, as far as I knew. He might, I suppose, have done it just as a sort of experiment, possibly feeling that the state had no right to say him nay. I don't know. In any event, my guess is that B-J pissed off his students by some snappy put-down and that's why they reported him. Given a good relationship between teacher and student, I doubt that they would have reported him, even if he had been guilty. As it was, they might well have reported him even if he wasn't. In any event, B-J was convicted, paraded in a humiliating 'perp walk' by the New York police and imprisoned at Eglin Air Force base in Florida. I recall seeing the odd book review from him in journals like *American Scientist*, bylined "Florida Rural Route..."

B-J always proclaimed his innocence. A sad postscript is that after he was released he apparently sent poisoned candy to the trial judge and a Duke colleague. He admitted that crime and was imprisoned again. Depressed, he stopped eating and died of pneumonia at the age of 67 in 1992.

Back to 1972. My first sabbatical leave was coming up. I knew where I wanted to spend it: the Zoology Department in Oxford, where there were several people whose work on evolution and animal behavior interested me. My host was David McFarland a lecturer and fellow of Balliol College. David and I shared theoretical and experimental interests. We were both taken with systems theory and cybernetics. In 1971 he had published a provocative book *Feedback Mechanisms in Animal Behaviour*. David had also done nice Skinner-box-type experiments on the control of feeding in ring doves. More on Oxford in the next chapter.

More broadly, I had become increasingly interested in adaptiveness, the evolutionary significance not just of instinctive behavior, a topic well-studied by others, but of learned behavior. Psychologists have tended to avoid this issue because it seems to involve final causes; it is *teleological*. Economists, in the most prestigious field of the social sciences, even now, despite its many manifest failures to predict, well... almost anything, are concerned with whether a behavior is 'rational' in the sense of maximizing money payoff. They largely ignore, or make wrong assumptions about, the actual process of reasoning. To say that some feature maximizes money or Darwinian fitness seems if not to miss, at least to skirt, the scientific point which is to understand causation. They maximize, yes, but how do they figure it out? Just how do they do it?

Biologist Colin Pittendrigh coined the term *teleonomic* for accounts like fitness-maximization which are in between purely teleological and causal accounts. His idea is that an explanation in terms of fitness implies a completely causal process of variation and selection, even though we can rarely supply the details, especially for selection that may have taken place in the remote past. Animal psychologists have a similar problem. They are very interested in *reinforcement* which is sometimes treated as a sort of final cause. Question: "Why does he do that (bark, scratch at the door, whatever)?" Answer: "Because he has been reinforced for it." The biologist's answer to a similar question about instinctive behavior is not dissimilar "Because it (the behavior) has been selected for" or "Because the behavior enhances evolutionary fitness". But in both cases, we believe that the teleological account rests on a causal process – evolution/learning via natural selection/reinforcement – even though the details may not be clear.

Top: **Jeff Bitterman, Pat Couvillon and Jennifer Higa, in Hawaii in 1992. *Bottom*: Elwyn Simons and friends at the Duke Lemur Center when he was director.**

A great virtue of looking at behavior as something that achieves a goal – or, in the jargon, optimizes something – is that it draws attention to environmental factors that are likely to be important. How to explain, for example, the 'irrational' timing effects I described earlier? Why does a pigeon respond in what looks like a rational way on fixed interval but not on response-initiated FI? On simple FI the bird acts as if he is balancing the cost of getting a reward later than necessary against the cost of making more responses than necessary. So he waits to respond after food, but not so long that, given that his timing is variable, not so long that he ever gets rewarded for the very first peck. Perfectly rational.

A response-initiated FI starts the fixed–interval timer not at the end of reward but after the first response. Clearly, the 'rational' strategy is to make that response immediately after food. Never happens: why isn't the bird 'rational'? Which is the wrong question because it presumes that 'rationality' is a thing, a kind of unitary faculty that you've either got or

not. In fact, of course, rationality, like every other adaptation, is the result of processes of variation and selection. And these are constrained, limited in different ways in different situations and for different organisms. In this case, the limitation is one of memory, the fact that the pigeon more easily remembers the time of the last food than the time of the last peck. Consequently, he uses the food, not the peck, as time marker – even though the peck is the event closest to the next food and so is the best time marker. As we saw, the food overshadows the peck which cannot therefore be used as a time marker for times longer than a few seconds. Looking at the rationality of the behavior in one situation and its irrationality in another similar one helps us understand the processes involved in both. But as an explanation, it is incomplete. So, on to Oxford
.

Chapter 12: *Benedictus benedicat*

We all moved to Oxford, to a row house, 19 Fane Road, at the end of the summer of 1973. Nick went to the local primary school; Jessica was still too young for school and stayed home with DB. I shared an office in the Zoology Department, in a modern building on South Parks Road which also housed Psychology. Fane Road is on the other side of the University Parks to South Parks Road. But to get through the park, you had to walk; cycling was not allowed. So I cycled a circuitous route via Marston Road through the center of Oxford to get to the Department each day.

Some people never learn. I bought a used car from a Duke colleague who was returning to the U.S. after a sabbatical leave at the University of Essex. It was an Austin 1100, basically the same model as the highly unreliable MG 1100 I took to Toronto in 1964. I assumed the problems I had then would have been fixed. Perhaps they were fixed, but new ones replaced them. The car was fine, except when it rained (this is England, remember), when it would stall. I had to cover the radiator with aluminum foil to keep the rain out. My colleague, a Durham neighbor and friend, was an anti-capitalist sociologist (hence the University of Essex). How many ill-omens does it take to get a young academic to make a good decision? More than two or even three, I guess.

Oxford is, socially speaking, rather like Paris in the summer. It is always full of (academic) visitors, a few distinguished most not – or at least not yet. So the locals are jaded, and not just in the summer. It is not a particularly welcoming place – unlike tourist-lite Paris in the winter when, in my experience, the locals are very kind and helpful. Unless there is some personal connection, Oxford academics tend to stay in their own circle doing their own thing.

There was something called the Newcomers Club, which involved visiting wives. DB joined that and got to know a lady whose brother was a master at Winchester College in Hampshire, an ancient public (i.e. private) boys boarding school: Motto: *Manners makyth man*. Founder: William of Wykeham; Date: 1382.

Winchester has always had a very strong academic reputation. We were sufficiently intrigued to accept the master's invitation to a tour later in the year, thinking possibly of sending Nick when he reached 13 in a few years. (We weren't thinking very seriously because Nick was only eight, because of the cost, and because of DB's reluctance to put Nick in a boarding school.) He pointed out the building where 'scholars' are housed. The way he spoke made this sound like a recent innovation, but when I asked when it had happened he responded 1394 or some date equally remote. The students lived in 'houses' run like a surrogate family. Even in 1973 there were no longer dormitories but rooms shared by I think no more than two students. Not exactly Dotheboys Hall – quite luxurious really.

The house master asked Nick a few questions. One was about sport and young Nick allowed that he played 'squash'. Confusion ensued when it turned out that Nick's 'squash' was not the adult game but a sort of poor-kids street game he had picked up at the local primary.

The other hurdle, should we consider Winchester for Nick, was the entrance examination. I still have copies of the exams for 1973. Topics are Divinity (example: "How did King Ahab acquire Naboth's vineyard?"), History and Geography, French, Latin I & II, Greek, Mathematics I & II, Science, Chemistry and English. All took between an hour and an hour and a half. They are pretty tough for an educated adult much less a twelve-year-old.

The first question in the English exam gave forty lines from "an Englishman shipwrecked in an imaginary country..." The last sentence reads: "Hence it follows of necessity that vast numbers of our people are compelled to seek their livelihood by begging, robbing, stealing, cheating, flattering, suborning, forging, gaming, lying, fawning, hectoring, voting, scribbling, stargazing, poisoning, libelling, and the like occupations; every one of which terms, I was at much pains to make him understand." Not exactly Harry Potter. The book is *Gulliver's Travels*, of course. The exam asked six questions about the piece, ranging from definitions of words like 'tracts' and 'avarice' to asking the student to explain various sentences.

My brother in law, Bob Rawlinson, had a friend who had been his high-school teacher and was now teaching at Stowe, another great 'public' school. Out of both curiosity and mutual friendship, I wrote to Frank Hudson about the prospects for Nick at Stowe. The news was not encouraging. Frank wrote that Stowe was "full" until 1977, three years ahead. And "we have 3,000 boys on the entry books, some of them for entry in the next decade." I was reminded that it was (and I suppose still is) customary among the English fee-paying classes to enter newborn males asap onto public-school waiting lists.

1973-4 was an eventful year in England. There was an energy crisis, both because of a jump in the price of oil and strikes by miners. From January to March 1974, the Conservative government, led by bachelor yachtsman and pianist Ted Heath, limited electricity to three days week. I can remember wandering around Blackwell's bookstore by candlelight – rather romantic, but hard on the eyes. There was a 'cod war' with Iceland (Iceland! Cod?) over fishing rights. In February 1974 there was an inconclusive general election and pipe-smoking Labour leader Harold Wilson (but he smoked cigars in private) took over a minority government. The IRA was also very active. There were bombings at Kings Cross and Euston stations that killed 13 people and two more at

Sloane Square and Oxford Street. The IRA campaign culminated 10 years later in the bombing of a Brighton hotel during a conservative conference in 1984. The target was Margaret Thatcher but the victims were five others killed, and many injured. The wife of conservative minister Norman Tebbit – a hero of my parents' (they had a signed picture) – was left permanently disabled. Islamic fundamentalists shouldered the terrorist burden in 2005 when four London sites, including Kings Cross, again, were bombed in a single day, killing 51 plus the four suicide bombers. (My wife and I had actually been at three of the four sites two weeks earlier.) "7/7" is now as iconic in the U.K. as 9/11 in the U.S. 9/11 was a bigger disaster, but England has suffered terrorism for much longer.

I began my year in Oxford quite sympathetic to the Irish. I grew up in Cricklewood, after all. But the Pavlovian pairing in 1973-4 of an Irish accent with repeated accounts of bloody barbarism conditioned me otherwise. Now it's hard for me to hear the brogue without a cloud passing.

David McFarland was a fellow of Balliol College and he got me a visiting appointment. It allowed me have lunch (free!) with the fellows and participate in various College events. Balliol, founded in 1263, has a distinguished history. Alumni include Adam Smith, Julian and Aldous Huxley, Richard Dawkins, Christopher Hitchens and Atul Gawande, not to mention conservative Canadian-American commentator Charles Krauthammer and five (as of 2015) Nobel laureates.

Balliol, despite its reputation as a 'modern' college, still retained many ceremonial remnants of its medieval origins. I remember having to wear a little gown (spares were hung outside the Common Room) to go to lunch, which was followed by tea or coffee – and possibly alcohol, but I was not a drinker so I don't recall. Another ritual, reflecting the Cheshire-cat fate of religion in England, was the vestigial grace said before meals: *Benedictus benedicat.* Still in Latin, but I daresay more was said in 1300, or even 1900. Even this threadbare remnant of Christianity has now disappeared from some Oxbridge colleges.

Quite a change from the eighteenth-century! Edward Gibbon, author of the magisterial *Decline and Fall of the Roman Empire,* in his autobiography described his first impressions thus:

A traveller, who visits Oxford or Cambridge, is surprised and edified by the apparent order and tranquillity that prevail in the seats of the English muses. In the most celebrated universities of Holland, Germany, and Italy, the students, who swarm from different countries, are loosely dispersed in private lodgings at the houses of the burghers: they dress according to their fancy and fortune; and in the intemperate quarrels of youth and wine, their swords, though less frequently than of old, are sometimes stained with each other's blood. The use of arms is banished from our English universities; the uniform habit of the academics, the square cap, and black gown, is adapted to the civil and even clerical profession; and from the doctor in divinity to the undergraduate, the degrees of learning and age are externally distinguished. Instead of being scattered in a town, the students of Oxford and Cambridge are united in colleges; their maintenance is provided at their own expense, or that of the founders; and the stated hours of the hall and chapel represent the discipline of a regular, and, as it were, a religious community. The eyes of the traveller are attracted by the size or beauty of the public edifices; and the principal colleges appear to be so many palaces, which a liberal nation has erected and endowed for the habitation of science. My own introduction to the University of Oxford forms a new aera in my life; and at the distance of forty years I still remember my first emotions of surprise and satisfaction. In my fifteenth year I felt myself suddenly raised from a boy to a man: the persons, whom I respected as my superiors in age and academical rank, entertained me with every mark of attention and civility; and my vanity was flattered by the velvet cap and silk gown, which distinguish a gentleman commoner from a plebeian student. A decent allowance, more money than a schoolboy had ever seen, was at my own disposal; and I might command, among the tradesmen of Oxford, an indefinite and dangerous latitude of credit. A key was delivered into my hands, which gave me the free use of a numerous and learned library; my apartment consisted of three elegant and well-furnished rooms in the new building, a stately pile, of Magdalen College; and the adjacent walks, had they been frequented by Plato's disciples, might have been compared to the Attic shade on the banks of the Ilissus. Such was the fair prospect of my entrance (April 3, 1752) into the university of Oxford.

But the promise was not fulfilled. Hierarchy and order, affluence and opportunity, were not enough. Gibbon concludes: "I spent fourteen months at Magdalen College; they proved the fourteen months the most idle and unprofitable of my whole life: the reader will pronounce between the school and the scholar; but I cannot affect to believe that Nature had disqualified me for all literary pursuits." Indeed...

I enjoyed for a while the modest ceremonies that remained in 1973 but then, after a couple of months, recoiled. They began to seem a bit silly. Yet they persist, many value them and they obviously serve a number of social functions, from promoting cohesion in the group to excluding outsiders. Do the good functions outweigh the bad? That's a topic for another time.

Top: **Ivan Pauliny-Toth and his wife, Jo, with DB at the Max Planck Institute for Radio Astronomy in Bonn in the early 1980s.**

But the free lunch drew everyone in and allowed, indeed forced, fellows to talk to scholars in different areas. Enclaves – 'silos' is now the term in administration-speak – were not encouraged. Cambridge ethologist Pat Bateson writes that he began his longtime collaboration with neurobiologist Gabriel Horne at just such a college meal. I got to chat with people like philosophers Rom Harré and Anthony Kenny – who succeeded Marxist historian Christopher Hill as Master a few years later.

My other memory of the Balliol Common Room is from a visit in 1985. A petition to block the traditional offer of an honorary degree to the then Prime Minister – Margaret Thatcher, an Oxford alumna – was being circulated. She was in fact denied the degree, an example of the pettiness of scholars that unfortunately is far from unique. I remember a rather similar and even more rancorous debate at Duke University in 1981 when the Presidential Library of Law School alumnus Richard M. Nixon was on offer. It could have been a great resource and the passion about Nixon would surely abate after a few years. But many faculty objected violently. Among the angriest was my colleague Norman Guttman, in all other respects a most amiable and reasonable man. Nixon was admittedly a far from admirable figure, but not without his good qualities and not much beyond the average of U.S. presidents, surely. Yet a vocal plurality of the faculty could not stand the thought of a 'Nixon Library' at Duke and persuaded their representatives to vote it down. There are tricky issues (no pun intended) in accepting an entity like the Nixon Library. Stanford University wrestled with a similar problem a few years later with the Ronald Reagan Library. But it would surely be better if the debate were driven by genuine concern for the integrity of the university, rather than by ideological animus. David Ferriero, current U.S. archivist and one-time Duke librarian, was asked recently during a talk in Durham about the Nixon library. He agreed that it was probably a mistake for Duke to turn it down.

Duke has often been idealistic, sometimes in odd ways. In the early 1980s, for example, a deal was almost struck to build the Whitehead Institute for Biomedical research at Duke. It would have been a great acquisition, but Duke was unwilling to concede any academic control to the Whitehead people. Good for Duke – I guess. But MIT was willing to go along with a more relaxed arrangement, and the Whitehead now resides there.

I got a few invitations to speak at English Universities and at meetings while we were in Oxford. One memorable occasion was at the University of Sussex, near Brighton, in the south of England. Sussex was the first in a flood of new universities that successive governments created in the 1960s and '70s, to be succeeded by magical conversion of almost all the pre-existing polytechnics to universities – obedient to the 'all shall have prizes' principle that had begun to take over education. After my talk – about which I remember little – we all went to a nice French restaurant. The host was Professor Stuart Sutherland, the chairman of the department. He had done interesting work on animal learning as well as in other areas. Stuart was brilliant but eccentric. Clad in bright yellow tie and what looked like a dark velvet smoking jacket, suitably flecked with ash, Stuart dominated the conversation. He discoursed on several topics, ending finally with one that stopped all side chatter. The issue was efficacy of a particular sex-toy he had bought for his wife from a shop in Tottenham Court Road in London (not far from UCL, actually). He listed in some detail the defects and problems he had returning the item. The show-stopper was that his attractive wife, Jose, was also seated at the table. I don't remember how the embarrassment of everyone was resolved – although the locals were less embarrassed than I because they had sat through similar scenes with Stuart before.

In any event, it turned out that Stuart was suffering from what would now be called bipolar disorder – mood swings from high, when he would write musicals to be performed on the London stage and the like, to depressive lows. That night was a 'high' obviously. He later sought treatment and wrote a devastating autobiographical critique of the nonsense that passes for psychotherapy, ranging from electro-convulsive treatment to the 'talking cure'. The book, published a few years after my encounter, is called *Breakdown* and it's still worth reading.

More out of curiosity than need, I applied myself for a chair at another 'new' university, York, that year. Fred Skinner even agreed to write on my behalf – damn nice of him, actually, given the swingeing critique of his 'superstition' experiment I had published just a couple of years earlier. I also asked Stuart Sutherland, since I needed as many English recommenders as I could get. But he declined because "I'm writing for one of my own people" – that would be I think Euan MacPhail at Sussex, who also became a friend, and in any case well-deserved the York job.

Back in Oxford, I found that David McFarland was not an easy chap. Brilliant, obviously, but not easy to talk to. He had a lovely wife, Jill, who worked for the Oxford University Press, and a son who turned out to be as odd as his father. They all lived in (here it is again) an odd modernish house in Cumnor, a village just to the West of Oxford. David has always been interested in the relation between animals, humans and machines and published several books on the topic. His most recent book (self-published through Amazon), is *Death by Eating,* on the effects of dietary change during human evolution. The last time I saw Jill was at an event held for David's retirement at Pembroke College, organized by Alex Kacelnik. Tragically, she was in a wheelchair, and apparently never recovered from a chronic ailment. David retired to Lanzarote in the Canary Islands and ran for a while some kind of eco-tourist B & B.

I did no experimental work at Oxford, although work continued at Duke and I did write up a number of studies. But the environment of the Zoology Department at that time was exciting. Nobel laureate Niko Tinbergen, whose *Study of Instinct* (1951) provided a framework for the study of animal behavior, was the professor. Niko came from an extraordinary Dutch family. Another brother, Jan Tinbergen won a Nobel in economics. (The economics Nobel sometimes presents a puzzle. In 2013, for example, the prize was shared by two people with incompatible views about the efficiency of markets - not that market efficiency is a well-defined notion. Evidently visibility, beats truth as a qualification.) Niko's third brother, Luuk, was also an ethologist and made some important discoveries to do with food selection in birds before committing suicide at the age of 39.

A leading intellectual influence was W. D. "Bill" Hamilton, who had come up with the core ideas of what was soon to be known as 'sociobiology'. He lectured during my year there but actually taught

Left: **At the Neurosciences Institute, Rockefeller University, New York in 1988.** *Back row:* **Bob Rescorla, John Krebs, JS, David Olton.** *Front row:* **Sara Shettleworth, Tom Carew, Alex Kacelnik, David Sherry, Pat Bateson.** *Right:* **Francis Crick at the opening of the new Institute after it moved to San Diego in 1995.**

(badly, judging by his lecture style) at Imperial College in London. His notion, that natural selection acts on 'selfish genes', was later brilliantly popularized by Richard Dawkins. One of Dawkins' papers, based on his PhD work, was on what he called a "peck, no-peck decision-making model", describing how chicks rewarded with food choose between stimuli associated with different reward values. It was straight operant conditioning, a line of work not so different from my own albeit with a variety of different species and, since non-Skinnerian, much more theoretical. Ethologists favored threshold-type choice models, but psychologists did not, until relatively recently. The details of these early models have not held up, but the idea that choice can be best understood in terms of a nonlinear competition among response tendencies, seems to be true. I remember looking at Dawkins' lengthy PhD thesis, which gave no hint, in either style or content, of the bestseller he was to publish a few years later.

Other faculty I remember from that time included Jeffrey Gray, a charming energetic and very open man who succeeded the famous H. J. Eysenck at the London Institute of Psychiatry several years later. Jeffrey and I certainly didn't see the world in the same way, but I enjoyed talking with him and respected his enthusiasms. Another psychologist was Michel Treisman, a quiet South-African-born psychophysicist, who came up with a pacemaker model that was later to be appropriated as the scalar expectancy theory of interval timing by animal researchers (see Chapter 10). His wife, cognitive psychologist Anne Treisman, soon achieved considerable eminence. She and Michel divorced in 1976 and she married Nobel-laureate-to-be Danny Kahneman a couple of years later.

Young ethologist John Krebs, son of the famous Hans Krebs of the 'Krebs cycle' in chemistry, had just graduated and left Oxford for the University of British Columbia. Efforts to get him back were underway and he returned to Oxford a year or two after I left. His *Behavioural Ecology* textbook with Nick Davies has been very influential. A handsome fellow with a deep voice (accent apart, he was pre-qualified to be the lead in a Mexican soap opera), he went on to a number of administrative roles, finally working for the government on food safety and health policy, where he has generally been appropriately cautious. But alas, like many technocrats, John is often willing to coerce people – er, I mean "guiding or restricting their choices" – for their own good. He became Principal of Jesus College and was styled finally Baron Krebs of Wytham (Wytham Woods being the Zoology Department field-research site near Oxford).

Alex Kacelnik, then a graduate student, now a faculty member, is a friend. He worked on decision-making in animals and a few years later demonstrated amazing intelligence in New Caledonian crows. The three of us have chatted at various times in meetings over the years. Ant man Ed Wilson, already famous for his monumental *Sociobiology,* gave Krebs, Kacelnik and me a tour around his lab at Harvard a few years later. Marian Dawkins, then wife of Richard, was doing interesting work on (for example) 'search image', about which more later. She has also made major contributions – real contributions, not just strong opinions or bad philosophy – to the thorny subject of animal welfare. Richard Dawkins' current wife is Lalla Ward, talented, aristocratic and comely one-time consort of Tom Baker as *Dr Who*. I remember she accompanied Richard when he gave a talk at Duke in the early 1990s.

In his autobiography Richard admires Duke's lemur collection, especially the little Aye-Aye, with its specially adapted middle finger which, as my Duke psychology colleague Carl Erickson showed, it uses to tap tree bark and thus detect grubs hidden underneath. Distracted by lemurs and new love Lalla, Richard placed Duke in South rather than North Carolina, but evidently made it to Raleigh-Durham airport and the plane to La Guardia.

I recall one strange incident several years later involving some of the Oxbridge people. John Krebs, Alex, Pat Bateson, (ethologist, head of the Madingley Sub-Department of Animal Behaviour and later Sir Patrick and Provost of Kings College, Cambridge), Tom Carew from Yale, who had worked with Eric Kandel on learning in the sea-slug *Aplysia*, I and a couple of others. The group consisted mostly of ethologists and animal-learning types. We were all invited by Gerald Edelman to what he rather cutely called an *Atelier* on theoretical neurobiology, at his Neurosciences Institute, then at Rockefeller University in New York City.

Gerry, who died in 2014, was a wonderful man: a 1972 Nobel laureate (for work on the immune system), a brilliant raconteur and a man of wide interests and sympathy. He was at that time working on his book *Neural Darwinism*, which came out a couple of years later. He and a colleague had devised a computer program to simulate some of their ideas about neuronal group selection in the brain. He invited everyone to come to the lab and see his simulation. I still can't believe the reaction of many in the group – a group assembled at Gerry's invitation and expense. Only Alex and I were willing to go with Gerry to see his new toy. I found out later that many of the U.K. contingent were upset at Gerry, who periodically appeared at the meeting unannounced and spouted off about his own ideas. I remember none of that because I always enjoyed hearing him talk. But still, to follow the *atelier* theme: *incroyable!* My last encounter with Gerry was a dinner after I introduced him as speaker to a conference of behavior analysts in Phoenix in 2009. He was as delightful and fluent as ever.

I had another memorable experience at the Neurosciences Institute a few years later. I was invited to give a formal one-hour talk at their annual meeting. I did not know it but I was apparently being vetted for election as some kind of associate. The associates were a distinguished bunch. At my talk were I think four Nobelists, in a group of not more than twenty or so: Torsten Wiesel, Francis Crick, David Hubel and of course Edelman himself. So far so good. The problem was as soon as I arrived in New York I realized that I had forgotten my slides (actual slides, of course, not a digital file). My wife had to courier them, at great expense and with some delay, to New York, which she did, narrowly avoiding a speeding ticket on the way. But my talk was stumbling. No offer followed. I think the honor (whatever it was) went to my competition Semir "Sammy" Zeki, a neurobiologist at UCL. Many years later I returned to the problem I spoke of so haltingly then, which turns out, I believe, to be much more important than I realized at that time.

During the Oxford year I worked hard to try and understand David McFarland's ideas about systems analysis and animal learning. He had what he called a 'state-space' theory but, in recollection, I don't think I ever fully understood it. My problem was relating the theory in a tight way to measurable data. I could not at first see how to relate the idea – so attractive when applied to, for example, the interaction between feeding and drinking – to something like operant choice behavior.

The major theme at Oxford at that time was of course evolution and adaptive fitness. A number of clever papers were being published showing, for example, that intermittent gliding in bird flight is not just avian fun, but often the most efficient way of getting around. Other 'fitness' questions: How many eggs should a bird lay to get the maximum number of offspring into the next generation, i.e. what is the optimal clutch size? *r* vs *K* strategies – whether 'tis better to have many young and neglect them (turtles) or just a few and nurture them (elephants)? *r* and K being the two parameters of the logistic (S-shaped) population-growth equation obeyed by an idealized population[1]. Feeding strategies: when to leave a food patch in a variable environment? Gordon Tullock, an economist, pointed out how utility maximization might apply to animal behavior. Tullock's 1971 note "The coal tit as a careful shopper" was a charming combination of wit and insight.

Do they maximize?

Much of this sounded to me rather like operant choice and soon after the Oxford year, I looked again at a classic operant choice problem, Dick Herrnstein's matching law, from this point of view. As usual the thing to be explained was not moment-by-moment patterns of response, but molar averages: the relations between average reinforcement and response rates on variable-interval schedules. The question was simple: is matching the optimal solution? Do pigeons maximize their rate of food intake by matching response and reinforcement ratios?

Solving the problem meant first of all finding something called a *feedback function,* that is, the mathematical function relating rate of response to rate of reinforcement on a given schedule. The feedback function for a ratio schedule is very simple: reinforcement rate is proportional to response rate. On FR 10, for example, if the pigeon responds 60 times a minute, he will get food 6 times a minute, and so on. Variable-interval schedules, the procedure used in matching experiments, are a bit trickier. But two properties of VI schedules point to a simple feedback function. Let's begin with VI 1 (60 s) and look at the two extremes. If the animal responds very slowly, say once per hour, he will get paid off every time – the 1-min interval timer will have set up for sure after that long a time. Similarly for once every two hours or even once every 30 minutes. So, the feedback function for very low response rates is the same as for fixed-ratio one: $R(x) = x,$ where x is response rate and $R(x)$ is the reinforcement rate produced by that response rate. At the other extreme, if the bird responds very fast, his payoff rate will equal the VI

---

[1] *K* is the maximum size of the population, *r* is its growth rate. A simple way to write the equation is in discrete-time form: $x(N+1) - x(N-1) = rx(N)(1-x(N)/K)$, which shows the change in population size, *x,* from step N to step N+1. When *x* is small, i.e., $x << K$, growth is exponential at rate rK; but as *x* increases, $K-x$, hence the increment at each step, decreases until $x = K$, the maximum population size.

rate. So, the feedback function must have an asymptote (i.e. must reduce to) $R(x) = x$, at very low response rates and to $R(x) = a$ (where a is the VI value) at very high rates.

The hyperbolic function, $R(x) = ax/(a+x)$, has these properties. If x is much less than a: $x << a$, the function reduces to $R(x) = x$, as x in the denominator disappears. If x is much greater than a, $x >> a$, then a disappears in the denominator and the function reduces to $R(x) = a$.

The math from this point is trivial, if you know elementary differential calculus[2]. First, set up some kind of constraint on total responding (otherwise the optimal solution will be to respond infinitely fast to both choices). Then compute partial derivatives $\partial R(x)/\partial x$ for the two choices, x and y, subject to the constraint that $x + y =$ some constant. The process finds the x and y values for which *marginal changes* are equal, that is, the values for which a small change in x (and thus y, since they must add to a constant) produces the same change in total reinforcement, which will be x and y values that maximize total reinforcement rate.

Table 12.1

		VI	Values			
	1		3			
x	R(x)	y	R(y)	x/y	R(x)/R(y)	R(x)+R(y)
19	0.950	61	2.859	0.311	0.332	**3.809**
20	0.952	60	2.857	0.333	0.333	**3.810**
21	0.955	59	2.855	0.356	0.334	**3.809**
30	0.968	50	2.830	0.600	0.342	**3.798**

The solution can also be found numerically – easy now with a spreadsheet. Look at Table 12.1. The two VI values are 1 and 3. The sum of responses, $x + y$, must equal 80 responses a minute. The left column shows 4 values for x, the other columns show the corresponding values for y, the two reinforcement rates, $R(x)$ and $R(y)$, and their ratios. The last column shows the total reinforcement rate: $R(x) + R(y)$.

The table shows two things. First, the maximum total reinforcement rate is indeed when x = 20 and y = 60, i.e. perfect matching. But second, deviations from matching don't make much difference. Matching is when $R(x)/R(y) = x/y = .333$, but even when x/y= .60 (bottom row), an 80% deviation from perfect matching, the total reinforcement rate declines by only 3 percent from the maximum. So it is very unlikely indeed that the matching animal is in some way adjusting its choice ratio until it achieves

---

[2] Staddon, J. E. R., & Motheral, S. (1978). On matching and maximizing in operant choice experiments. *Psychological Review, 85,* 436-444.

the maximum payoff rate. Only if the pigeon responds very slowly, a response or two per minute, say, does his preference make much difference to his overall payoff rate.

In subsequent years, many attempts were made, in my own lab and by others, to see if overall reinforcement rate can exert any kind of direct control over response rate. A strong test is the following *frequency-dependent-payoff* schedule, explored by my student John Horner for his PhD dissertation[3]. It works like this: the pigeon has two choices. Payoff probability is always the same for each choice. But the value of that probability depends on the relative frequency of the two choices. For example, in one procedure, payoff probability (for both keys) was zero if the animal only chose left. It was maximal if he was indifferent, but low again (though not zero) if he chose only right. Choice probability was computed over a 'window' of 80 past choices[4].

Obviously if the animal can adjust his choice to maximize payoff, his choice proportion should peak at 50:50. He should just alternate. No pigeon did so. Instead all showed a bimodal pattern of preference, with peaks on either side of the maximum – sometimes they preferred the left sometimes the right; they were never simply indifferent. So, even in a procedure where a change in molar choice proportion makes a big difference to total (molar) payoff rate, the pigeons fail to adjust optimally.

What were they actually doing? We don't know in detail, but broadly speaking the animals seemed to be reacting to each reward with a burst of pecks on the rewarded side. Several rules specifying exactly how they do this give similar bimodal preference distributions[5].

Another experiment asking this same question – does average reward rate control average response rate? – looked at just a single response. We used a modified ratio schedule whose value decreased as a function of postfood time: the longer the animal waited to start responding the fewer pecks he needed to get the next bit of food[6]. By varying the rate of decline and the maximum ratio size we got a range of feedback functions. The question: did the pigeons adjust their behavior so as to maximize payoff

---

[3] Horner, J. M. Probabilistic choice in pigeons. Ph.D. Dissertation (Duke, 1986); Staddon, J. E. R., & Horner, J. M. (1989) Stochastic choice models: A comparison between Bush-Mosteller and a source-independent reward-following model. *Journal of the Experimental Analysis of Behavior*, **52**, 57-64, Figure 1.

[4] 32 choices in some experiments; the window size made little difference.

[5] Staddon, J. E. R. (1988) Quasi-dynamic choice models: melioration and ratio-invariance. *Journal of the Experimental Analysis of Behavior, 49*, 303- 320.

[6] Ettinger, R. H., Reid. A. K., & Staddon, J. E. R. (1987) Sensitivity to molar feedback functions: A test of molar optimality theory. *Journal of Experimental Psychology: Animal Behavior Processes, 13*, 366-375. This is called an *interlocking* schedule.

rate as the molar feedback functions changed? The answer again was an unequivocal "no".

Conclusion: whether the actual behavior is optimal (reinforcement-rate maximizing) or not, the molar (average) variables used to compute the optimum play no causal role. Instead the animal reacts to each situation with a set of heuristics or rules that work well enough – especially in situations similar to those its ancestors lived through and were naturally-selected by – schedules like fixed-interval. But they don't behave optimally on schedules like the response-initiated FI schedules I described in the last chapter for which there is no counterpart in nature, at least nature as experienced by ancestral pigeons.

Top: **Staddon lab circa 1980: John Horner, Ken Steele, JS and John Hinson.**
Bottom: **JS and Alliston and Leonor Reid on their wedding day, March 26, 1983, in Cheraw, South Carolina.**

Two things seem to have combined to exaggerate the importance of response rate: the steady response rate that animals develop on VI schedules, plus Skinner's emphasis on response *probability* as the critical datum. But the pattern on VI, just like the pattern on FI or on a spaced-responding schedule, is an adaptation – good in the case of VI and FI, not always good for spaced responding – to the contingencies of reinforcement. The VI pattern has no special significance. The assumption that VI response rate = response probability = Skinner's response strength is a theory or at least the outline of one. But because of Skinner's bias against theory, it was unexamined. A little thought might have revealed early on that a single number is not adequate to describe an organism's responding, its *state,* even under the tightly controlled conditions of operant schedules.

During the late 1970s and 80s my laboratory acquired a number of wonderful graduate students, postdocs and research associates who contributed creativity as well as effort to the research. In addition to Nancy Innis, Ginny Simmelhag and Mike Davis: John Malone, John Kello, Bettie Starr, Alliston Reid, John Hinson, Richard "Chip" Ettinger, Ken Steele, Susan Motheral, Jennifer Higa John Horner, Derick Davis, Mark Cleaveland, Sandy Ayres and Kazuchika Manabe, studied different aspects of the general problem: how do animals adapt to reward schedules: what do they do and how do they do it?

Most of my students went on to academic careers, but a few renegades defected, usually to good effect. Karen Kessel, who did her PhD on altruism in lemurs (the only time I was ever directly associated with work at the Duke Lemur Center), went on to be a successful consultant. Art Kohn who, embarrassingly, now calls himself a cognitive brain scientist – such are the imperatives of marketing – taught at several places on the West Coast but then went on to start his own teaching company. Art loves teaching and while a student on his own hook taught a no-credit undergraduate class that attracted quite a few students. Qualified as a baseball umpire, Art persuaded me to attend my one and only Durham Bulls game. John Penney, onetime DJ and road manager for the jazz group Spyrogyra, was bright and capable. But not, as it turned out, all that interested in psychobiology. He lasted only a short time in the lab and returned to music.

It was always my strategy – at first unconscious but later, when I saw what I was doing, quite conscious – to allow a new student to follow his own intuition in designing a project. If that worked out, fine. If not, if the student's project failed or looked as if it was going nowhere, I would work with him (or her) to come up with a joint project, probably based on an idea of mine.

It wasn't until the mid-1980s that I became convinced that the overwhelming emphasis on molar response and reinforcement *rates*, a key part of B. F. Skinner's legacy, was not the best way to understand the processes that underlie operant behavior. In the meantime, I tried to make sense of the extensive molar data on reinforcement schedules. A clue was provided by the work of David Premack. He is the inventor of what has come to be called the *Premack Principle* (PP). Working with monkeys, which have a range of things they like to do, even in an impoverished captive environment, Premack proposed (a) that behavior itself is/can be reinforcing, and (b) that access to behavior A can reinforce behavior B if, under free conditions, A is more probable than B.

A grand theory

The Premack Principle is very general. It assumes that all behaviors are equally reinforcible (given the right conditions) and all can, in turn, serve as reinforcers. This is not always true. Many activities – excretion, salivation, other Pavlovian reflexes – can't be conditioned at all by operant means. Even some skeletal activities, like wing-flapping by pigeons, are reputed to be hard to reinforce with food (apparently the problem can be overcome under certain conditions) and many other 'constraints on learning' have been amply documented. The Premack principle has been taken by many to mean it is the activity not the object that is reinforcing – the eating not the meal.

But this also is not universally true. People seem to value art and other luxury items for the object itself, not the activity associated with it. Heroin addicts value the 'high' not the injection. On the other hand, the PP suggests that many 'operant' activities, like a monkey fiddling with a flap or a lever in his cage, can be used to reinforce one another. Premack showed that if fiddling with a lever occurs more frequently than looking through a hole, the frequency of hole-peering can be increased if 10 s of hole peering are necessary to get 10-s access to the lever.

Figure 12.1

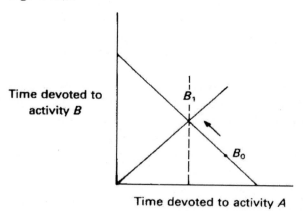

Time devoted to activity A

The idea that there is a preferred level (probability, rate, proportion of time taken up) for each activity suggested a simple generalization. Imagine first of all just two activities, A and B (Figure 12.1). The two activities take up the whole time. This constraint is represented by the diagonal line: $A + B = $ constant. The preferred levels (times taken up) of these two under free conditions can be represented by a point, B_0, on that line in this two-dimensional space. (In the example preferred *time A* > preferred *time B.*) Suppose, like Premack, that we make one activity dependent on the other, according to a one-to-one (FR 1) schedule such as 10 s of *A* gets you 10 s of *B*. This schedule can be represented by a straight line of unit slope through the origin in the space. Given these two constraints and two dimensions, there are no degrees of freedom left. When the schedule is imposed, time taken by activity *A* must decrease and time taken by *B* increase, as the new point representing the state of the two activities moves from B_0 to B_1.

This increase in *B* looks like the kind of thing that Premack found: *A* takes up more time than *B*, *A* depends on *B*, apply a 1:1 contingency, and *B* increases. But obviously the change in the level of B is entirely due to the time constraint; it is what is called a *restriction effect*. It's not an effect of A as a consequence, which would be a reinforcing or *contingent effect*[7].

This argument shows that there is a problem with Premack's interpretation of his results. Imposing a schedule constraint, like fixed-ratio 1, between a more-frequent and a less-frequent activity may yield an increase in the less frequent for two reasons: restriction and/or reinforcement. To get anything more than restriction, there must be at least three activities, so that Figure 1 expands to a state space (echoes of McFarland here) of three (or more) dimensions. The diagonal constraint line in turn changes to a plane (or a hyperplane, for more than three dimensions). Now imposing a schedule constraint between activities A and B, where A takes up more time than B, may produce changes in the levels of all three activities: A will decrease, B and/or C ('other' activity) may increase to fill the total time. Let's suppose, as before, that B increases. How much of that increase is due to a reinforcing effect and how much to restriction? To answer that question, we need some kind of theory. One possibility is that there is no reinforcing effect at all, so that when A is tied to B, and given that under free conditions A > B, then the time freed up by the constraint is filled by increases in B and C proportional to their free levels – as if the three activities were three balloons squashed into a box. If one balloon is further compressed, the others will expand proportionately.

Another possibility, which is one that I explored in some detail, is that the changes in activity levels under constraint represent an adaptive, homeostatic process[8]. Suppose that the distribution of activities under free

---

[7] Note for economists: the distinction between restriction and contingent effects is exactly the same as the distinction between *income* and *substitution* effects.

[8] Staddon, J. E. R. (1979). Conservation and consequences—theories of behavior under constraint: An overview. *Journal of Experimental Psychology: General,*

conditions, B_0, is like a regulatory set point, so that the organism will adjust his behavior so as to be as close to it as possible: formally, so that the total *cost of deviations* from B_0 is a minimum. It should be possible, then to predict the effect on the operant response (B in the figure) of different values of the reinforcement schedule. The example is fixed-ratio one, but obviously other FR values, and other schedules − like variable-interval − are possible. If we know the feedback function, and we know how distance from B_0 is measured in this hypothetical Euclidean space, we can derive, for each schedule type, a function relating steady-state response and reinforcement rates − the usual experimental measure of reinforcement effects.

There are two aspects to the cost of deviation from B_0: the metric, linear or some other, for the displacement between the point representing each behavior and the free-behavior value, B_0; and the *relative costs* for different activities. Presumably the cost of deviations from your normal eating rate is much greater than the cost of deviation from your normal rate of lever-pressing or key-pecking. Indeed, a low-CoD behavior is, almost by definition an operant (conditionable-by-consequences) behavior.

When a high-CoD behavior is contingent on a low-CoD behavior, and the *x* vs *R(x)* points are plotted, the result is what is called a *response function*. I compared the response functions the model predicts for ratio and interval schedules with the empirical data. The empirical functions are sufficiently variable from animal to animal that I did not expect to match them exactly. But I was happy that the *minimum-distance model* easily duplicates the *differences* between the two schedules. The function for ratio schedules is mostly sloped negatively, which shows a basically regulatory response: as the ratio increases, so does response rate, partly compensating for the drop in reward rate that would otherwise be caused by an increase in ratio value. Conversely, the function for interval schedule has a generally positive slope, representing an increasing shift away from 'other' behavior as the interval schedule delivers more frequently. When the reinforcement rate is the same, response rate is generally higher for ratio than interval schedules. For all these features, the model matched the data.

Search image

Another idea from behavioral ecology that caught my attention was the concept of *search image*. The German ethologist and philosopher Baron von Uexküll in the 1930s reported the following experience:

108, 1-3. Staddon, J. E. R. (1979). Operant behavior as adaptation to constraint. *Journal of Experimental Psychology: General, 108*, 48-67.

When I spent some time at the house of a friend, an earthenware water pitcher used to be placed before my seat at luncheon. One day the butler [we're talking barons here!] had broken the clay pitcher and put a glass water bottle in its place. When I looked for the pitcher during the meal, I failed to see the glass carafe. Only when my friend assured me that the water was standing in its usual place, did various bright lights that had lain scattered on knives and plates fly together through the air and form the water bottle.

This led von Uexküll to the concept of *search image*, which is a sort of perceptual learning. You (or an animal) repeatedly experience some stimulus – like a new prey species. At first, you don't notice it, but then, suddenly you do. Perhaps this idea accounts for the fact that when a new prey type – a differently marked bug, for example, first appears in an area it is relatively free of predation. But after a while, the birds 'catch on' and the intensity of predation increases. What's going on? Can the predator really not 'see' the new bug? Does it experience the kind of epiphany that von Uexküll reported? Or, less anthropomorphically, is there such a thing as perceptual learning in animals? And what exactly does that mean? Do they actually become better at *detecting* a new prey item? Or do they just take a while to learn that it *is* a prey item?

Marion Dawkins and others did several clever experiments to see which of these hypotheses is correct. For example, suppose we scatter two different-colored but equally palatable grains on a plain background and measure the proportion of each that a pigeon eats. Usually he will peck at them in proportion to their density: eating 80% of type A and 20% of type B if A and B are in the ratio 4:1. Now repeat the experiment with a background that makes type A more difficult to see than type B. What will happen then? Well, probably the pigeon will first pick mostly type B, but then, as it learns it will switch and concentrate on higher-density A.

Two things seem to be going on. First there is some evidence that the birds are learning to recognize A against the confusable background. Their ability to actually *detect* A improves. And second, they may be adjusting their search *speed* in order to do so: going more slowly so as to see and eat the more abundant but harder-to-see type. Rob Gendron, a biology graduate student and I showed this effect in an operant-conditioning setting[9].

---

[9] Gendron, R. P., & Staddon, J. E. R. (1983). Searching for cryptic prey: The effect of search rate. *American Naturalist, 121*, 172-186. Staddon, J. E. R., & Gendron, R. P. (1983). Optimal detection of cryptic prey may lead to predator switching. *American Naturalist, 122*, 843-848. Staddon, J. E. R., & Gendron, R.

Search image is real and has real human implications. For example, bicyclists are at great risk of being hit by cars in places where cyclists are rare. Once in a while, a motorist will fail to process his visual image correctly and will simply not 'see' the cyclist. A fatal accident may be the result. In places where cycling is common – many towns in Europe, for example – car-cycle accidents are rare. In North Carolina, as I can attest from the number of friends and acquaintances who have been hit, they are much more frequent – although as more people now cycle, that may be changing.

The claim that a motorist may simply not see a plainly visible cyclist might seem implausible. But a few years ago a couple of social psychologists did a very clever experiment that showed just this effect: if attention is diverted, even a very obvious stimulus may be missed. The situation was this. Subjects watched a short video in which two teams of three, in black and white shirts, are tossing a basketball. Subjects are told to count silently the number of passes made by the white team. Then, after a minute or two, the video stops and the subjects are asked "Did you see the gorilla?" And indeed, a girl in a black gorilla suit, waving her arms about, walked behind the players in the middle of the movie. The gorilla was on screen for a full nine seconds. But only about half the subjects report seeing it[10]. Distraction like this is the secret of very many magic tricks. Attention can be diverted in ways of which we are completely unaware.

Theory of mind

David Premack, whose principle got me started on the line that led to the MD model, is an interesting figure. On the one hand, his original principle is behavioristic in the extreme, a sort of blank-slate learning law. On the other hand, he was working mostly with monkeys, and later chimps and human infants, and soon became interested in more cognitive aspects of their behavior. Premack fell right off the behaviorist wagon when, with Guy Woodruff, he proposed what he called *Theory of Mind.* In a 1978 paper they wrote:

> In this paper we speculate about the possibility that the chimpanzee may have a "theory of mind": one not markedly different from our own. In saying that an individual has a theory of mind, we mean that the individual imputes mental states to himself and to others (either to conspecifics or to other species as well). A

P. (1983). Search image and the optimal detection of cryptic prey. *Proceedings of the 18th International Ethological Conference* (p. 269).
[10] http://www.theinvisiblegorilla.com/videos.html

system of inferences of this kind is properly viewed as a theory, first, because such states are not directly observable, and second, because the system can he used to make predictions, specifically about the behavior of other organisms.

The choice of words was brilliant. *Theory of Mind* caught on, big time, and their paper has been cited thousands of times. Here's a human example. You show a 3-year-old a Crayola box. "What's in the box?" you ask. The kid says "Crayons." Then you open the box and show that it's actually full of candles. The next step is to introduce a third party previously not present. A cuddly Snoopy doll will do. Question to the child "What do you think Snoopy will say if I ask him what's in the box?" "Candles" the child replies. A five-year-old, on the other hand, will reply without hesitation: "Crayons", revealing perhaps his superior understanding of the knowledge possessed by a naïve observer.

Results like this are interpreted as showing a fundamental change in the child's view of the world. At three he thinks that everyone is the same; but at five his world is much more nuanced. In the words of a well-known developmental psychologist: "By the time they're five they have a view of the mind that is much more like ours. They understand that things can be tricky and deceptive. That you can change your mind. That things aren't always the way that they seem."

This account seems entirely reasonable to us, as adult human beings. But it makes a giant leap. From the simple failure of the 3-year-old to switch his response, to the successful switch by the five-year-old, we are generalizing to…what? a world view? a view of human nature? The "imputation of mental states"? This is empathy, not explanation. There are, after all, much simpler possibilities. Maybe the 3-year-old can only remember one response at a time. He just learned the box contains candles: box = candles. That's all he has room for, so he answers a question about the box with the response he's already learned. You could test this, maybe, by systematically testing his memory. Or by waiting for 10 or 30 seconds between the demonstration and asking the question. The delay will usually cause earlier memories to win out over newer ones. Or maybe the child thinks that Crayola now makes candles, so he should forget his previous belief that Crayola = crayons; perhaps he hasn't learned that such a distinctive box can contain different things. Another possibility is something to do with stimuli. There are two stimuli: box and Snoopy. Maybe 3-year-olds can't do conditional discriminations: You (the experimenter) mean "candle" but he (Snoopy) means "crayons". You could test this in ways that don't involve people. These tests might be done with 3-year-olds and 5-year-olds to see if they differ in the suggested ways. None of them says anything about "mental states" or imputing them to others.

Easy to say; hard to do. There are many possible hypotheses for the results of a pretty complicated experiment. It's hard to test any of them adequately, especially with kids. We can't give them too many tests; they will get bored or annoyed (who can blame 'em!), and their parents will object after a while. But, though feeble, these suggestions, at least hint at potential *explanations* for the phenomenon, explanations in terms of elementary processes we can study independently. Jumping from Crayola and box to a child's 'theory of mind' really explains nothing. What *is* 'mind' that the child – or we – can have a theory of it? On the other hand, the observation of a simple performance difference between 3- and 5-year-olds is certainly interesting. Enough differences like that might suggest better hypotheses than my rather simplistic suggestions. So, more observation, please; less leaping to plausible-sounding but largely meaningless conclusions.

Here's a theory-of-mind example with chimps from Premack and Woodruff[11]:

> Rather than confronting a chimpanzee with an inaccessible object and observe his possible problem solving, we have instead shown him a human actor confronting inaccessible objects, and asked the animal to indicate how he thought the human actor would solve his problem. Specifically, we made four thirty-second videotapes of a human actor in a cage similar to the chimpanzee's struggling to obtain bananas that were inaccessible in one of four different ways: They were (1) attached to the ceiling and thus out of reach overhead, or (2) outside the cage wall and thus horizontally out of reach, (3) outside the cage, but with the actor's reach impeded by a box inside the cage, located between him and the bananas; in the last case (4), not only was the actor's reach impeded by a box, hut the latter was laden with heavy cement blocks.
>
> In addition to the four videotapes, we took still photographs of the human actor engaged in the behaviors that constituted solutions to the four problems. In one case, he was photographed stepping onto a box; in a second case [lying on his side and reaching out of the cage with a rod]; in a third case, moving a box to the side, and in the last case, removing cement blocks from a box.

---

[11] Premack, D. & Woodruff, G. (1978) Does the chimpanzee have a theory of mind? *Behavioral and Brain Sciences, 4, 515-526.*

A comprehension test for the animal (Sarah) consisted of showing her each of the videotapes in turn, putting the last five seconds of the tape on hold, and then offering her a pair of the photographs, one constituting a solution to the problem and the other not.

Sarah was a very experienced, 14-year-old chimp. She had watched much TV and solved numerous reasoning problems over her life. She was rewarded in this experiment first verbally "Good Sarah, that's right" vs "No, Sarah, that's wrong" and then, at the end of the session, with a food treat. She chose the picture with the correct solution almost 100% of the time. After discussing and dismissing two alternative accounts of this rather extraordinary behavior, P & W conclude: "The view we recommend... is that the chimpanzee solves problems such as the present one (and others a good deal more complicated) by imputing states of mind to the human actor." Hence, 'theory of mind'. Again, this is a suggestion that relies on human empathy for its persuasive power. Indeed, P & W comment: "it is important to note that empathy and 'theory of mind' are not radically different views; they are in part identical" although they are talking about the animal's empathy with the actor, not our empathy with the 'theory of mind' account.

But there are alternatives – tough to test, as with the toddler example – but surely worth considering. Sarah's task is simply to pick the right picture from the pair. Any characteristic of the *state* the movie sequence and the picture induces in her that predicts reward will do the job. (I use 'state' as a summary for the repertoire of potential actions that is available in a given context, as I described earlier.) The most obvious characteristic is to choose the picture that matches what *I*, Sarah, would do in this situation. Never mind the human actor. This works, even to the extent that Sarah did worst on the problem that ape-intelligence pioneer Wolfgang Kohler found the most difficult, removing the cement blocks. Since apes find that task the most difficult they are presumably most unlikely to recognize the solution to that problem.

Apes and toddlers are quite intelligent. Even in an age of advancing AI we still know very little about human intelligence. Yet imputing 'theory of mind' to another organism takes for granted that we do actually understand how human beings solve problems like these. We see an ape do it and conclude "Look, he has a theory of mind just as we do". But that surely begs the scientific question which is always 'How does it work? in us as well as them. Exactly what processes are involved?' We understand a cognitive process when we can in principle build a machine, real or simulated, to perform the task that is evidence for it. Saying that the chimp or the child has a 'theory of mind' not only misses that target it isn't even aimed at it.

It's not too harsh to say that 'theory of mind' is much like the theory of 'intelligent design' (ID) as an alternative to Darwinian evolution. ID advocates claim that a structure like, say, the flagellum of bacterium *E. coli*, is 'irreducibly complex', in the sense that it is impossible to imagine how it could have formed via stages each of which is adaptive. The idea is that only the final structure makes sense; all possible intermediate structures are useless. It's like a mousetrap, which doesn't work until all the bits are in place. So how could it have evolved through a progression of bits? Well, maybe some things are 'irreducibly complex' – or maybe we just aren't creative enough to imagine the right bits or the right selection environment that might have produced them? In any event the core problem with intelligent design is that it doesn't explain the designer: How does *he* work; how does he figure out the best way to make a flagellum? So also with 'theory of mind': Great, but what *is* a mind? How does *it* work?

Nevertheless, we owe David Premack a debt for coming up with a number of surprising experiments and some fruitful and provocative ideas. I last heard from him in 2008. We had quite a lengthy correspondence. The social coin of the realm for scientists is not money but priority (think Mayr and Greenewalt, mentioned earlier). Gregor Mendel gets much more credit for genetics than de Vries, Correns and von Tschermak who independently re-discovered it 34 years later. The first guy to publish wins. Apparently the credit-assignment process failed for David Premack at least once. The high-prestige, hard-to-get-published-in journal *Nature* published a paper in 2007 on how infants evaluate social behavior. David felt that the paper failed to acknowledge, plagiarized even, a mass of work that he and collaborators had published over the previous 17 years. Premack tried to get *Nature* to give him proper credit, but the editor was unresponsive. The correspondence shows how keenly scientists feel any failure of their field to acknowledge their priority. I am not qualified to adjudicate. I have not studied a vast and generally unconvincing literature in developmental psychology. But I suspect that DP has a point.

In 1981 I was asked to give a talk as the Initial Plenary Lecture of the XVIIth International Ethological Conference in Oxford. The title was *Psychological aspects of animal economics*, and I spoke about the minimum-distance model – not the most riveting topic for a Conference more used to discussions of the mating habits of hyenas or the displays of bower birds. I was, of course, nervous and, in an era before PowerPoints, it was easy to run over time. I did, and got heckled for my pains by Robert May, a man I much admired as a pioneer in the study of mathematical chaos theory. (He showed, for example, that even the simple completely deterministic logistic growth equation – see footnote 1 – with certain parameter values can behave in an apparently chaotic way.) New Zealander May (subsequently Baron May of Oxford, Scientific Adviser to the U.K. Government and President of the Royal Society) was of course quite right to complain. When I mentioned the incident 30 years later in a letter to the *Financial Times* commenting – favorably – on a piece of his he responded kindly, saying that he did not recall the incident. Well, he wouldn't would he?

Chapter 13: Nuoro

In 1980 I edited a collection of chapters for a book called *Limits to Action: The Allocation of Individual Behavior.* The topic was optimization in behavioral psychology, biology and economics. The animating idea was that optimization theory could provide a unifying principle for all three areas. I was able to get contributions from a botanist, three biologists, two psychologists, and three behavioral economists. With one exception all dealt with nonhuman research, which is perhaps why this branch of behavioral economics has been rather neglected in favor of the human variety, represented by stars such as Richard Thaler, Danny Kahneman and the late Amos Tversky. Expanding on this theme, in 1983 I put together a large book, *Adaptive Behavior and Learning*[1] which was somewhere in between a monograph and a textbook. It was an attempt to apply both optimality ideas and more real-time dynamic ideas to the full range of data on animal learning – all in the context of evolutionary biology. The book was well-reviewed and got an unexpected boost a few years later from Nobelist Gerald Edelman when he turned his interest from the immune system to behavior.

In the late 1980s I got a letter from someone called Professor Giulio Bolacchi at the University of Cagliari. I had no idea where Cagliari was (it's the capital of Sardinia), but apparently Prof. B was interested in translating *Limits to Action* into Italian and would I write a foreword. I wrote the foreword and heard no more.

Then in 1990 or so my secretary Vanessa said someone with a foreign accent had phoned to offer me a teaching stint in Italy. "Veery interesting!" I thought, mentally adopting a suitable accent. I contacted the caller, who was Italian, but working in the U.S. He was in fact calling from Wisconsin on behalf of Prof. Bolacchi who had set up some kind of graduate business school in Sardinia. Prof. Bolacchi had previously worked on the Board of RAI, the Italian public television system and apparently had enough influence with the Italian government to get money to set up a rather visionary organization: AILUN: Associazione Istituzione Libera Università Nuorese, which being translated stands for The Free University of Nuoro. Bolacchi's admirable intention was to boost the poor economic situation of Sardinia by training smart kids in ways that might improve the country. AILUN was a tiny graduate school which offered master's degrees in two areas: The Science of Organization and Applied Optical Technology. (In later years, a course on tourism was added.) An odd mixture, although Italy does have a wonderful optical tradition,

---

[1] Staddon, J. E. R. (1983). *Adaptive behavior and learning*. New York: Cambridge University Press. Pp. xiii, 1-555. Reprinted, 2009, 2nd edition 2016.

beginning, I suppose with Galileo – but not, of course, in Sardinia. The Science of Organization program reflected Prof. Bolacchi's intellectual interests. He is a sociologist who believed in the unity of science; hence the attraction of my book, which brought together economists, biologists and psychologists.

Sardinia and AILUN: Maria Assunta and Rosanna in front of a Nuraghe. *Insets:* Rita (top) Milena (middle) and Paola.

AILUN was (and still is, in a less international form) in Nuoro a small town in what was then a rather primitive region in the Barbagia region in the center of Sardinia. Bolacchi's plan was to teach the master's course mostly in English with the aid of foreign – mostly American – professors. The course lasted six months or so and each professor would teach for four days. On the fifth, Bolacchi would appear, coming up from Cagliari where he was a professor at the university. He would review material with the students and have them produce little projects. The course was rather rigorous, especially once the process settled down after a few years. Students would often work late into the evening to produce a presentation for the next day.

So I agreed to lecture for a week. I found out that three other contributors to *Limits to Action* were also on Bolacchi's list that first year. Getting to Nuoro was not easy, until traveler-hostile but route-friendly

Ryanair started a direct flight from the U.K. to Alghero, on Sardinia's west coast. But this first time I had to travel via Rome and then on an unreliable airline called Meridiana to Cagliari in the south. I was met at the airport by an elegantly dressed young woman with a delightful Italian accent who introduced herself thus: "Allo! I am Maria Assunta Vinci, and I am 'ere to take you to Nuoro." Which she did in a nice Mercedes at extremely high speed. The two-hour trip took us through increasingly primitive-looking places – elderly women, heads covered, clad entirely in black, would gaze at us with looks that went from indifference to puzzlement.

Cindy and JS having dinner with Giulio Bolacchi and Rosanna in Su Gologone, our hotel near Nuoro.

Francesca Bolacchi meets us near St. Peter's, in Rome.

In later years, AILUN bought its own car (A Fiat not the Benz, which belonged to Maria's physician husband) and had its own driver, one Sebastiano Satta – same name as a famous Sardinian poet. Mr. Satta was no poet; he was an emotional man and on one occasion, obviously very upset about something, he drove us at truly hair-raising speed to Olbia to catch a plane. It's the only time I have really feared for my life in a taxi. My guess is that Mr. Satta was a political appointment: AILUN must have hired him as a favor to some person of influence. He certainly didn't get the job for his driving or his English skills.

Nuoro nestles on a saddle between two mountains, but I was to stay at a nearby village called Oliena at wonderful hotel, Su Gologone, which means 'spring' in the Sardinian language. It did have its own spring, in a beautiful grotto right below the hotel. It was and is a delightful place, nestled under a mountain in the Supramonte. The décor matches the environment, with original oil paintings by Sardinian artists in every room. The warm and comely Proprietress Giovanna and her little daughter Camilla completed the picture. The hotel was surrounded by a fence and there was gatehouse at the entrance. *Carabinieri*, the police, were frequently in evidence. The significance of these features was to become clear a year or so later.

The first year, teaching was rather haphazard. Prof. Bolacchi seemed to have a pool of charming young women whose duty it was to attend to visiting teachers. So, that year I had lunch with Paola and Milena in picturesque restaurants, either downtown on a nearby mountaintop. In later years Rita Olla became the mainstay, running the course. The lunches, still delightful, were in Nuoro, in a restaurant called *Canne al Vento*, named after a novel by Sardinia's most famous writer: Nuoro's Grazia Deledda. Young Rosanna Farci, who began as just a camera operator (the classes were all videoed – a legacy of RAI, I suppose), became over the years Bolacchi's chief assistant and driver (he never drove himself). One of the great pleasures of my visits was the dinner, usually on a Thursday evening, with Bolacchi, and Rosanna – who had learned very good English and became a real colleague.

The classes were fun although the students were quite variable in their ability to speak English. Discussion tended to be dominated by the one or two who spoke English well. But the class was always small: 10-15 kids about 50:50 male and female. The foreign teachers were not required to grade assignments (what a relief! Grading – marking – has always been the most hateful part of teaching to me). Grading was handled by Bolacchi and Rosanna. Which was great, but did mean that feedback was limited. I could never be quite sure how much I was getting across. This improved in later years when the final day was taken up with presentations by groups of students.

In the second year, I was taken to see part of the island – Roman ruins in the south and the *Nuraghi,* distinctive circular towers twenty or fifty feet in diameter, 4000-year-old remnants of the ancient Nuragic civilization and found only in Sardinia. I noticed only one on my first visit, but saw many more later. Hundreds of them are scattered all over the island.

Top: **Nuoro, seen from Oliena.**

Bottom: **Signora Bolacchi (second from right) and daughter Francesca in the Signora's Rome apartment. Cindy and JS on left.**

In October 1991 I was in Nuoro at the same time as the Clarence Thomas Supreme Court confirmation hearings in the Washington. The sexual innuendos and other distasteful behavior attributed to Thomas made the hearings of international, albeit prurient, interest. The young women of AILUN had apparently been following the hearings and I expected the same negative reaction that Thomas elicited from academic women in the U.S. Well, no. "It's just what some men do" they said "What's the problem?" The allegations were never substantiated and Thomas, amid much controversy, was eventually confirmed.

After a year or two my wife would accompany me and we would take a few days after the course to holiday in Rome or Florence. Our first time in Florence we stayed in a hotel on the Arno river recommended by a much-traveled Duke colleague. Nice, but very American-tourist, we thought. On our second day there, the TV was on and suddenly I saw a familiar sight: the mountain behind Su Gologone and in front of it a policeman speaking about some crime. Well, it turns out that Italy has a fine tradition of kidnapping, with the Supramonte region in the lead, since the mountains are excellent cover. In 1979 two anarchist rock musicians, Fabrizio De André and boyfriend Don Ghezzi were kidnapped and held captive in the mountains near Oliena. They were released only after a ransom was paid. A few years earlier Jean Paul Getty III, the 16-year-old grandson of American billionaire J. Paul Getty was kidnapped in Rome by a the 'Ndrangheta', a Calabrian mafia. A ransom was paid only after his parents got a parcel containing some hair and part of the poor kid's ear (not that the boy was an angel – many thought at first that he had set up the 'kidnapping' to extract money from his rich but penurious grandfather). In 1992 an eight-year old boy was kidnapped in Porto Cervo, in the north of Sardinia. This caused general outrage – even a plea from the Pope. So kidnapping was in those days an Italian tradition – a tradition with its roots in Sardinia.

Porto Cervo is on the Costa Smeralda, the posh area in the north where the Aga Khan owned 35 miles of coastline and the rich parked their yachts. We rented a small apartment there one rather cool Spring. The most exciting thing to happen then was that after a few days, a stray cat (we'd been feeding her), came and presented us with her kittens. Rather sad, since we couldn't take them.

We learned later that the TV in Florence was about yet another kidnapping in the Barbagia region: a 65-year old lady and a young woman. That was all we knew. When I returned the next year I was naturally curious about what had happened to these people. Well, as best I could discover, the older woman had been killed and the young one had gone off with one of the kidnappers. The general belief was that the whole thing had been rigged as a cover for the romantic liaison. Tough on the older lady of course. And the body? Again, the story was that when kidnapped people were killed – which happened quite a lot – the body would be cut up and fed to the wild boars which roam the island – Sardinian recycling…

So now I had a better idea of the reason for the *carabinieri* at Su Gologone, the surrounding fence, the gatehouse, occasional police dogs and so on. I asked our host later whether we, visiting lecturers, were protected by anything other than our relative poverty. Well, yes, he assured us. The proprietors are local people, they know the criminals and have made a suitable arrangement with them. Which somewhat reassuring.

Another year we took our holiday in Rome, in a small hotel near the Spanish Steps. We got to meet Prof. Bolacchi's gracious wife and daughter Francesca. They lived in an eighth-floor apartment overlooking St. Peters and the Sistine Chapel. In the living room was a full-length oil portrait of Signora Bolacchi and next to it, a picture by Joan Miro. We all had lunch with Bolacchi's sister and her husband in an outdoor café nearby. Although his wife lived in Rome, Prof. Bolacchi of course lived much of the time in an elegant apartment in Cagliari, but the arrangement seemed to work for them. Daughter Francesca, who very kindly met us near St. Peters where we gave her a large bunch of flowers, has gone on to become a biomedical researcher.

Chapter 14: Dahlem

In 1980 or so I got a letter inviting me to something called the Dahlem Konferenzen. The topic was Animal Mind/Human Mind and the number of participants was small, just 13. The guiding force for this and many succeeding conferences was an impressive and elegant lady called Silke Bernhard, a medical doctor by training. Dahlem conferences were interdisciplinary. The structure of the four-day conferences was rather rigid: *Deutsche Ordnung*. Speakers reported on the state of one of four sub-fields, not just their own work. The contributions of each of the four groups were summarized by a *rapporteur*. People circulated comments and questions during the meeting. A book of the commentaries and *rapporteurs'* reviews for each conference was published by the Foundation. In 1983 I attended a second meeting, this one on The Biology of Learning[1].

The meetings were held in the Dahlem district of Berlin, on the 19th floor of the glass-modern Europa Center, opposite the ruins of the *Gedächtniskirche,* a bombed church left as it was in 1945 as a reminder of WW2. As always in German conferences of this sort (at least in my experience) entertainment was provided in the evenings. The bit I remember was a performance of an *opéra bouffe*, something by Offenbach (possibly *La Grande-Duchesse de Gérolstein*). I cannot recall the name (he wrote many, many operas and operettas), but I do recall a scene with dallying swains and maidens, a meadow and a battlefield in which huge cannons, shaped as giant upturned phalluses, were wheeled across the stage. The audience of respectably dressed Germans sat through it impassively; no one cracked a smile.

I remember a similar mixture at a von Humboldt event near the *Tegernsee* in Bavaria later. On the evening of day one the entertainment was provided by four attractive young girls, primly and properly clad in dark skirts and starched white blouses: a string quartet, playing Mozart. Very nice. The next night, appeared several middle-aged men, tummies bulging over leather shorts and suspenders (braces to you Brits), who sang Bavarian drinking songs while cracking whips and jumping up and down on the tables. It was great fun, but the contrast was a bit head-snapping.

I'm still not sure even now how I feel about the highly structured Dahlem format. At the second conference, I remember one of the *rapporteurs*, an ornithologist specializing in the behavior of songbirds, was required to report on a session much of which was devoted to classical conditioning. This was cruel and unusual punishment. Classical

---

[1] 1981 March 22-27, *Animal Mind - Human Mind* D. R. Griffin (ed.). 1983 Oct 23-28 *The Biology of Learning* P. Marler, H. S. Terrace (eds.)

conditioning is an arcane, quantitative, methodologically complicated and not very biological area. The ornithologist bridled at the esoterica – both mathematical and procedural – and at one point looked as if he could not continue. It was one inter-disciplinarity too far, at least for him!

Germany *Top:* **Cindy and Juan and Ute Delius in Konstanz.**
Bottom left: **JS and 'Krampus" during tour of Bavaria.**
Bottom right: **Bavarian entertainment – on a von Humboldt trip to Rottach-Egern.**

The meetings certainly accomplished their objective: summarizing just what is known about an area. They mapped the known and 'known unknowns'. But the fixed format meant there were few opportunities to encounter 'unk-unks' – unknown unknowns, things not suspected at the outset. For this, less structure not more is needed.

So for me the main virtue of the meetings was the people. I met some old friends from the U.K. and made new ones from other places. Marion Dawkins, John Krebs and Norbert Bischof were at the first meeting; Pat Bateson, Mark Konishi, Klaus Immelman, Peter Marler, Stephen Lea and Juan Delius at the second.

Berlin was still, in 1983, a divided city. Pat B and I explored East Berlin. At the heavily guarded checkpoint we bought with dollars or pounds our obligatory 25 marks. East Berlin in 1983 was a strange experience. On the one hand was the incredible Pergamon Museum with wonderful Assyrian structures like the Gate of Ishtar and the Pergamon

240

Altar. But for the fact that they were in East Berlin, and before that in Nazi Germany – eras when criticism of any official action was an invitation to trouble – surely removing those wonderful ruins from their native country would be regarded by many as simple theft? (Even the legally obtained sculptures brought from the Parthenon to England in the early 19th century by Lord Elgin have come under fire.) But what would those critics say now when ISIL is deliberately destroying Palmyra and beyond? Whatever the ethics, the Pergamon was a great experience.

On the other hand, my general impression of East Berlin was that it was like bomb-free parts of London just after the war. The streets were strangely quiet. The only sound was walking feet, because traffic was light to non-existent. The people looked grim. The streets were free of parked cars, just as I remember them in London. Shop signs looked archaic (mind you, in Germany many still do). We were in a posh part of the city and checked out the fancy stores, the East German equivalent of Fortnum and Mason. The comparison was utterly pathetic. Huge windows contained just a handful of goods for sale. The shops looked quite bare. I seem to recall that many left-wing (and even a few right-wing) economists at that time thought the GDP per capita of East Germany was higher than Italy's. Perhaps they were the same people who thought the Soviet Union was a great economic success and were surprised by its collapse in 1991. In any event, a comparison of E. Germany with almost any Western nation would find East Germany on the bottom, as the failure of the Soviet bloc demonstrated.

Pat and I took a train, to spend some of those otherwise useless marks. When we got out we found we could not return to West Berlin, except through our original entry point.

Silke Bernhard's role in the Dahlem Conferences ended in 1989. I don't know the details, but apparently she was unwilling for the conferences to lose their autonomy by being merged with a larger – and doubtless more bureaucratic – entity. So she retired, to science's great loss. The conferences resumed, under the management of the Free University of Berlin, in 1990.

One of the people I met at the Dahlem gatherings was Norbert Bischof, a German, but at that time professor in the University of Zurich in Switzerland. He had an interesting background, beginning with the great ethologists Konrad Lorenz and Erich von Holst. Norbert had spent some time at Caltech and we shared an interest in systems theory. In his later years he became interested in human morality and the incest taboo. He invited me to give a talk in Zurich and assured us (my wife came too) that we had slept in the same bed as chimp pioneer Jane Goodall, in the little apartment the university reserved for visiting notables. This experience was only topped a few years later when I stayed with Terence

Kealey, then Vice-Chancellor of the University of Buckingham. As England's only private university, Buckingham was proud to have as it Chancellor (titular head) Margaret Thatcher, who had also stayed in Kealey's house.

German art. *Top:* **In Konstanz Peter Lenk's 30-foot rotating statue 'Imperia'.**
Bottom Left: **In Heidelberg.**
Center: **In Aachen.**

Norbert B lived and worked in the German-speaking part of Switzerland, but apparently the relations between Germans and Swiss was not always harmonious. The Swiss among themselves spoke *Schweizerdeutsch*, a dialect comprehended only with difficulty by Germans. Even though most Swiss speak standard German, apparently they often revert to *Schweizerdeutsch* when they wish not to be understood by a German-speaking foreigner.

Zurich is the epitome of spick and span respectability. I ordered coffee in a restaurant which fit the image, with an immaculate white-haired, dirndl-clad *frau* in charge. But it is the first and only time I have found a cockroach – yes, that's right, a cockroach – in my coffee. What was interesting was the reaction of the *frau*. She was unapologetic and clearly thought the whole thing was my fault.

My other association to Zurich is an incredibly depressing book called *Mars* by Fritz Zorn, given to me by Katharina, a German friend. *Mars* is an autobiographical account of growing up in Zurich by a young cancer victim who died at 32. Zorn blamed Zurich society for his affliction:

> I am young and rich and educated; and I am miserable, neurotic, and lonely. I come from one of the very best families on the east ("right") shore of Lake Zürich, also known as the "Gold Coast." I had a bourgeois upbringing, and I have been a model of good behavior all my life. My family is fairly degenerate. It is likely that I have much genetic damage, too, and I am maladjusted. Needless to say, I have cancer.

Zorn's book made quite a splash when it was first published – in Germany – which it tells you something about how Zurich then seemed to Germans.

The most important person I met in Dahlem was Juan Delius. A fine scientist and an open and enthusiastic character he has an interesting history. Born and raised in Argentina, his father was apparently a manager on the family estate of another famous Argentinean psychobiologist Fernando Nottebohm, then at Rockefeller University. Juan did his doctoral work in Oxford under Niko Tinbergen. Then he taught for five years in the University of Durham, in the north of England, before taking a position in the Ruhr University in Bochum. It was through Juan that I got an award from the Alexander von Humboldt foundation that allowed me to spend a sabbatical year in Bochum. Juan and his wonderful wife Ute had many friends in Durham who greeted them warmly on their occasional visits. It was shocking to me that he was not permitted to visit the U.K. after the Falklands War began in 1982.

The Humboldt Prize came at a critical juncture in my personal life. DB and I had been struggling for many years. I will not go into the sources and symptoms, except to say that they were probably more apparent to others than to us. Relatively recently, I had a conversation with a Berkeley colleague who had been at Oxford at the same time as DB and I. Apparently we had him and his wife to dinner in Fane Road. Now he asked how I was and at some point I told him that DB and I had been divorced for many years. "I knew there was a problem" he said. Apparently it was obvious, even in 1974.

When the Humboldt offer appeared, DB said that she was not willing to come to Germany with me for that year. I was puzzled, since she liked Europe and was usually keen to travel. She had started a little computer business – *Microglyphics* (The *Countess Ada Lovelace*...book was a byproduct of my work on a database program for the company) in which Nick had also been involved – but that didn't seem to be the reason for her reluctance. Other family ties were slipping: Nick had already left home and my daughter was in her last year or so of high school. So, quite quickly, DB and I split; I left McDowell Road and rented a little apartment. At almost the same time, a longtime family friend, Lucinda, was having her own marital difficulties. Cindy is a poet and short-story writer, but her day job was as head of collections for the Durham County Public Library, an amazingly good library for such a small town as Durham was in those days. I had known Cindy for many years without any romantic involvement. Indeed she was much more DB's friend than mine. Her daughter had been a baby-sitter for Jessica more than once. Her two children were also out of the nest and soon she left her husband and found her own place. Cindy and I were companionable. Pretty soon we were together and have remained so ever since. Divorce was the worst experience of my life up until that time; but eventually, the legal horrors were over. Cindy and I were married a couple of years later, with my parents as happy witnesses, in Aldershot (still 'Home of the British Army' despite some dilution in recent years).

Cindy, at age 48, had never been outside the U.S. This was soon to change, big-time. First, in the Spring of 1986 I had been invited to give a series of lectures in several different places in Brazil, a three-week excursion. More on Brazil later. Then, in the Fall, I was to begin my one-year sabbatical in Juan's lab in Bochum, Germany.

So in late summer 1986 Cindy and I, our luggage and a cooler of giant water bugs (*Lethocerus*, if you're interested) arrived in Frankfurt to catch a train to Bochum. The water bugs were a favor to a Duke friend, Michael Reedy, who studies how myosin works in muscle. We were delivering them to Mike's German colleague who met us at the airport, collected our luggage, put it all in a huge trolley which he proceeded to take on an escalator – which surprised us but they've made it work in Frankfurt – and was an enormous help negotiating the trip from the gate to the train station, all in the giant Frankfurt airport. Would the *bugs* be allowed on a passenger flight nowadays? I wonder…

Bochum is in the Ruhr region of Germany, which was devastated by Allied bombing during WW2. By 1986 it had been pretty thoroughly rebuilt. Nevertheless, most Germans express sympathy when I tell them where we spent my sabbatical year. But we liked Bochum. It has a strong theater tradition, apparently but, alas, we never learned enough German to appreciate it. But it does have wonderful walks. We enjoyed going to the nearby Kemnader See and to a little restaurant at the top of a grassy hill not too far away. There was a vast indoor swimming pool, the Hallenbad Querenburg, roughly the size of Lake Baikal. Shopping was close by even if the shop attendants in the local Karstadt were more like the Zurich *frau* than our friendly black ladies in the Durham Kroger. And there was an excellent little restaurant, *Die Uhle* (The Owl) downtown. I am not a huge fan of German food, but the menu at *Die Uhle* was memorably excellent.

There *was* a general food warning when we arrived in Germany. Avoid *wild*: meat like venison or wild boar. The reason was Chernobyl, the nuclear disaster which occurred on April 26 1986. The fear was that wild animals might be radioactively contaminated.

We bought a great used car, a 1980 Mercedes 280E, old enough to be 'grandfathered' back to the U.S. without having to add too much in the way of catalytic converters and the like. But we rarely used it. One memorable time, not having a clue where we were going, we drove into the middle of the large pedestrian plaza in the center of Bochum. We got out without incident and successfully navigated the multilayer subterranean car park underneath.

We lived in a guest house for foreign visitors, endowed by the Volkswagen Foundation. On the other side of the valley was a factory of their competitor Opel. Waste heat from the *Opel Werk*, was piped to many residences around the valley, including our guesthouse. Opel went bust in the new millennium. I wonder what local housing now does for heating? We were the only English speakers; Stephen Lea, from Exeter in England, came later, in the middle of the year, also to work with Juan. We acquired a couple of parakeets (budgerigars); I named them Henry and Min, after the querulous old couple in the old BBC radio Goon Show (starring Peter Sellers and Spike Milligan among others). The birds were no trouble – until someone started talking, at which point they would join in enthusiastically. They were a gift from a newly married graduate-student couple. The birds were owned by the guy, one of Juan's students, but disliked by his wife-to-be, hence the donation. (I felt that a more dedicated student of animal behavior might have ditched the fiancée). One of the guesthouse residents was a lively, highly patriotic – jingoistic, even – physicist from Pakistan, proud to be working on nuclear weapons. Our main contact there was the very nice cleaning lady we referred to as Frau Staubsauger (the German word for vacuum cleaner), until Cindy chatted to her and learned that her real name was Trudy.

The Ruhr University is in a dozen or more multistory 1960s concrete tower blocks. Apparently in 1986 it held the German record for suicides, although whether that was a morale problem, or just because of the many high-altitude opportunities it offered, I don't know. I was assigned a large office on the ground floor, which slightly surprised me. As a visitor, I had expected something more modest. The reason soon became clear: when it rained, the ceiling tended to leak. As the office was on the ground floor of a building of ten or more storeys, this was a puzzle. It wasn't a real problem, but it showed that the construction quality of these rather ugly 1960s buildings was not what one would expect from Germany.

The University arranged for subsidized touristic trips for visiting students and faculty. These involved a bus, which, in our case, took us to the southern parts of Germany around Christmas time of that year. The trip was hosted by a formidable middle-aged lady called Frau Zeher. We were the oldest passengers and also the most senior. Frau Zeher therefore gave us the royal treatment. We were also the only native English-speakers, which led to some interesting translation problems. The trip was delightful. We enjoyed the *Christkindlmarkt* in Munich, various museums, and the *Krampus* monsters in a cafe. The *Krampus* is a horned figure who disciplines misbehaving children at Christmas time, a sort of anti-St. Nicholas – a much needed figure in the contemporary United States, I think.

Juan had a lively lab of a dozen or more people. There was one further addition. While I was still at Duke in the Spring of 1986 I got a letter from Clive Wynne, a young Brit interested in our timing work, which had been the starting point for his PhD thesis at the University of Edinburgh. Did I have funds that would allow him to come to Duke and work with me as a postdoc? I wrote back and told him I would be in Juan's lab that year not at Duke, but he was welcome to come there if he could get the money. This was all good, it turned out, because Clive also wanted to work with Juan and he was able to get money from a European source.

Dynamics of timing

But the real point of our trip was science. At the center of my interests at that time was – *time*. Earlier work on timing and the omission effect led me to a rather grand hypothesis: that time-to-food is all that matters to the hungry pigeon. Much, if not all, their behavior on reinforcement schedules, may be explainable in this way. Clive and I decided to test this idea in several ways. So we set up some experiments in Juan's lab. One experiment[2] worked like this. Two conditions were to be compared. In the *fixed* condition the first peck after food reinforcement changed the key color from red to green, and food was delivered after T s in green; then the cycle repeated. Time to the first peck, t, was under the control of the animal. We measured the steady-state value of t as a function of several fixed values of T the delay-to-food in green. This function, t vs T, was compared with the same function in a second condition, the *dependent* condition, where T was not constant, but depended on the value of the preceding wait time t. In fact, in the second condition, $T = K/(t+1)$, where K was varied from one phase of the condition to another. Again, we measured the steady-state function relating t to T.

---

[2] Wynne, C. D. L., & Staddon, J. E. R. (1988) Typical delay determines waiting time on periodic-food schedules: static and dynamic tests. *Journal of the Experimental Analysis of Behavior, 50*, 197-210.
http://dukespace.lib.duke.edu/dspace/handle/10161/3387

To understand this experiment just think about the animal's optimal strategy. What should he do to get food as soon as possible? In the fixed condition, the answer is simple: respond as soon as possible, so t should be close to zero and independent of the fixed delay T. We knew from other work that this was unlikely to happen, but the conditions here were a little different, so we couldn't be sure. For the dependent condition, however, t should depend on K. For example, if $K = 10$, the best value for t is 2 s, whereas if $K = 20$, the best value for t is around 8 s.

None of these expectations panned out. In every case, t bore the same straight-line relation to T, i.e. wait time was proportional to time-to-food, as measured from the previous food, and the proportionality was independent of the schedule. A second experiment confirmed these results. In other words, under very many conditions, pigeons that get food at periodic intervals are *obliged* to wait a fixed fraction of the interval before they begin to peck – no matter what the dependent relation between delay and wait time and no matter what is best for maximizing food rate. We called this process *linear waiting*.

A third experiment showed just how powerful linear waiting is. We again adjusted the relation between time-to-food and waiting time. The logic of the procedure can be laid out in syllogistic fashion as follows:

1. The interfood interval equals wait time, t, plus a time T set by the experimenter.
2. Time T begins with the first post-food peck, as in the previous experiments.
3. *Pigeon:* Suppose that a pigeon is obliged to wait for a time equal to 50% of the preceding interfood interval, e.g. $t(N+1) = .5(t(N)+T(N))$, where N is the number of the interval. This is obligatory linear waiting, based just on the one-back interfood interval (this is rather a strong assumption, but as we'll see it is true under many conditions).
4. *Schedule:* Suppose that post-peck time to food, T, in interval $N+1$ is proportional to wait time in interval N; i.e. $T(N+1) = kt(N)$, where k is a parameter that is varied from condition to condition.
5. This little system then boils down to two simultaneous difference equations:

$$t(N+1) = .5(t(N)+T(N))$$

and

$$T(N+1) = kt(N).$$

It's easy to see that if proposition 3 is correct, if the pigeon really adjusts his wait time to be a fixed fraction, like .5, of the previous interfood time, then this arrangement sets up what is called an *unstable equilibrium*, like a ball balanced on top of an upturned bowl. For example, if k, the schedule parameter, is equal to 1, the situation is always stable. Just pick any pair of initial values for $t(N)$ and $T(N)$ and plug them in to the two equations: 1 and 1, stable result: 1; 1 and 2, stable result: 1.67; 1 and 3, stable result: 2.33, and so on. The ball balances on the bowl. But if $k < 1$, then t gets smaller and smaller from cycle to cycle, converging on zero. With the initial conditions 1, 1, if $k = .5$, then successive wait times are .5, .375, 3.125 and so on getting down to .000125 after 30 cycles. The ball falls off one side. Conversely, if $k > 1$, t increases from cycle to cycle without limit. For example, if $k = 1.5$, then successive wait times are: 1, 1.5, 1.875…to 82 after 30 cycles. The ball falls off the other side.

So if our hypothesis is correct, this experiment should show some pretty weird, and rather improbable results. In particular, if k is greater than a critical value, the pigeons should wait longer and longer after each food delivery and, perhaps finally quit responding at all. This seems highly improbable. Why should the bird quit responding when all he needed to do was wait a short time after each food delivery before starting to peck? Well, rather to our surprise this is exactly what they did. We changed the value of k from day to day and each day the wait times either increased or decreased depending on whether k was greater or less than a critical value. Linear waiting, the idea that pigeons' wait time adjusts from interfood interval to interfood interval seemed to be confirmed[3].

This little experiment is one of the very few examples of a real dynamic analysis of pigeon learning. I hoped to see more research along these lines. But the logic is a little tricky and most research has focused on the measurement – *ad nauseam* – of well-understood static properties, in situations contrived for the orderly results they yield rather than the light they shed on underlying processes.

'Transitive inference'

As understanding of elementary processes both advances and becomes more technical, collective attention drifts. Psychology is to most people the 'science of mind', after all. So, when the going gets tough, attention tends to revert to the conventional. People increasingly want to understand the ways in which animals are like people: Are they conscious? Do they have empathy? Language? A theory of mind? Interest in *comparative cognition*, as this movement was called, was gaining traction in the late 1980s. In Juan's lab, less affected than most, but not insensible of the

---

[3] Some of the details are still open, of course. Wait time depends on the preceding interfood interval, but effects of earlier intervals – and even relatively remote history, yesterday, last week – under some conditions are not ruled out by this single experiment.

trend, the focus was on something called *transitive inference*. The analogy was to human logic: if I tell you that A is bigger than B and B is bigger than C, and then ask you: "Which is bigger. A or C?" you have no problem answering: "A". The question was: Can animals do something like this? My concern was: suppose they can: how do they do it?

Juan and his collaborator Lorenzo von Fersen, used with pigeons a simple procedure that had been used to show (it was claimed) transitive inference in monkeys and children[4]. The birds were presented over and over, over several days, successively with four pairs of easily distinguished visual stimuli, one on each pecking key. The payoffs were as follows, where "+" means the bird got some grain for pecking the stimulus and "–" means that he got a few seconds timeout (lights out, no food). (Stimulus pairs were occasionally presented without rewarding either one, just to see what the birds had learned.):

A+B–
B+C–
C+D–
D+E–

This is a tricky procedure for the pigeons, because the relationship between stimulus and payoff is inconsistent: stimulus B, for example, is rewarded when paired with C, but not when paired with A.

The birds performed pretty well – as well as human infants and chimps – even though three of the five stimuli, B, C and D, are both rewarded and unrewarded, depending on the pairing. Accuracy varied from close to 100% at the extremes, A+B– and D+E–, to 70-90% for the pairs in the middle: B+C– and C+D–.

The key test was then to present a new pair, never seen before: BD. Would the animals follow transitive logic and choose B over D? Bearing mind that B and D had both been reward and not rewarded, so (the assumption was) based on reward history alone, neither should be preferred. The pigeons chose B over D with quite high accuracy: 80% or so, better than some training pairs. Bingo: 'transitive inference' in pigeons, now shown to have a capacity possessed only by 'higher' animals like children or apes.

But is this really transitive *inference*? Well, it certainly isn't the kind of thing that allows human beings to order items according to logic. It occurred to me that there is an asymmetry in the training series that might underlie the result. The end stimuli A and E have extreme values: A is

---

[4] Fersen, L. von, Wynne, C. D. L., Delius, J. D., & Staddon, J. E. R. (1990) Deductive reasoning in pigeons. *Naturwissenschafften, 77*, 548-549.

always reinforced, and E never is. If these values 'rub off' somehow on stimuli with which they are paired, and so on down the line, each stimulus will acquire a value in the desired order: B more value able than C (even though equally often rewarded and not) because paired with the highest value stimulus A, C more valuable than D (even though equally often rewarded and not) because paired with a stimulus more valuable than the one paired with D (B > C), and so on. We called this *value-transfer theory* and suggested it as an explanation for transitive inference in 'lower' animals. A few experiments in other labs seemed to confirm VTT.

The problem, of course, is that VTT is ad hoc. It does serve to show that pigeons do not need to reason logically to rank-order the stimuli as they did, but it introduces a completely new principle with no support other than the fact that it works in this situation. It's just not parsimonious.

Honeybees suggested an answer. Bees are a wonderful corrective to what one might call 'theory of mind' thinking, the ascription of human-like abilities to animals. Bees, for example show something called 'metacognition', a fancy name for knowing what you know – as when asked the capital of Estonia you either say it, say you don't know, or say you're pretty sure you know it, just give me a minute. But no one suggests that bees engage in conscious reflection. Nevertheless, metacognition has been convincingly demonstrated in bees and a few other non-human species. It was also known that bees can solve discrimination problems quite similar in structure to the TI procedure.

Pretty soon, a notoriously crusty old-tyme comparative learning theorist, M. E. "Jeff" Bitterman and his longtime collaborator Pat Couvillon, both experts in bee behavior, showed that a simple learning model could simulate both the bees and v. Fersen's result. The model also does a good job of predicting the accuracy of each discrimination: highest at the extremes, and different for BC vs CD. They chose the oldest and simplest learning model, something called Bush-Mosteller after the two guys, a psychologist and a mathematician, who first proposed it. B-M is incomplete and fails to account for many elementary properties – memory effects, reversal learning and so on. But Couvillon and Bitterman made their point: even though the TI procedure seems to equate payoffs for the three middle stimuli, the process animals use to make the discriminations means that the *values* of those three stimuli (and of course the two end points) will not be the same – so in preference tests they will show the apparently transitive ordering they did. It turns out many other learning models also work. Pretty clearly, "a complex conditional inference process can be modeled by processes which themselves are neither conditional nor inferential."[5] Once again, this little story shows the importance of getting to the level of the learning *process*, how the animal's state changes from moment to moment. Without some understanding of this, no matter how crude, a number like percent correct is impossible to interpret.

The Humboldt Prize allows for one or two repeat visits to your German host. I wanted to visit Juan's lab again, but in the meantime he had moved, to the University of Konstanz. The reason was an odd Harvard-type rule applying to academic promotions in German universities, all of which are, of course, state institutions. Apparently, after a certain point, a professor cannot be promoted within his own institution: he has to move. So it was with Juan: to move up the ladder he had to get an appointment at another institution – in this case, Konstanz.

His replacement in Bochum was one of his own students, Onur Güntürkün, known affectionately as "the Turk on wheels" because he was confined to a wheelchair. Onur's specialty was neurophysiology and behavior and he has gone on to a highly productive career. In 2003 he achieved his fifteen minutes of media fame for an observational study of how people kiss each other in railway stations.

Konstanz is a beautiful small town on the shore of Lake Constance (in German the *Bodensee*). Zurich, the closest airport, is about 40 miles away. Across the river connecting the Bodensee and a smaller lake, the *Untersee*, is the twin town of Kreuzlingen, in Switzerland. The first things you see when you walk into Kreuzlingen are banks, many of them.

We spent a few months with Juan in Konstanz on two occasions. Surprisingly, we didn't like Konstanz quite as much as Bochum, despite its physical beauty. There were pleasant walks around the lake which offered occasional surprises. On one occasion, chatting animatedly with a friend, I suddenly noticed that the people sitting on the grass all around us were completely naked. Strict in many things, the Germans are relaxed about nudity, dogs (allowed in restaurants) and even, to some degree, about smoking, the great modern sin – at least compared to the U.S. Nudity is tolerated, even encouraged. College girls would strip to the waist at the first sign of sun in the Spring. Many public saunas are co-ed and nude. But Konstanz is basically a tourist town and, topless coeds apart, the university seemed a little less lively than the Ruhr University.

Public art in Konstanz and many other German towns is very good. Most of the sculpture is figurative, some a bit grotesque or sexual (large penises seem to be a favorite, although whether this reflects wish-fulfillment or lack of imagination I can't say). But odd or not, the art –

---

[5] Wynne, C. D. L. (1995) Reinforcement accounts for transitive inference performance. *Animal Learning and Behavior, 23*, 207-217.

usually sculpture – is always skillful and interesting. None looks like missile debris or something left behind on a construction site. Most impressive of all in Konstanz was a 30+-ft statue on a tall plinth at the harbor entrance. It was by artist Peter Lenk and was apparently erected clandestinely on private ground without municipal permission shortly before we arrived. It is of *Imperia*, a prostitute who has in her outstretched hands two miniature figures representing 15th-century eminences: Pope Martin V and the Emperor Sigismund. The statue, which rotates slowly on its pedestal, was inspired by *La Belle Impéria,* a story by Honoré de Balzac, which satirizes the corruption of court and church. When we were there, there was a petition in a little shed at the foot of the statue and visitors were invited to sign and say whether they approved or disapproved of *Imperia*. She's still there so I guess most people liked her. We certainly did.

German art has less of an international reputation than it deserves, I think. Katharina, a friend in Regensburg, a beautiful medieval town on the Rhine, introduced us to Albrecht Altdorfer, a 16th century artist comparable, I would say, to any renaissance painter, but unknown to most English-speakers.

Possibly our slight disappointment with Konstanz was because we could not get into the university accommodation for visitors but had to rent a rather inconvenient 5th floor walk-up downtown. I confess I did get a little testy when I found that another American visitor on another occasion did manage to get the good digs. Why, I wondered? Well, it turned out he had *bribed* someone at the university. Which reminded me again – although the visitor, a psychologist colleague, was an exception – that Americans, and American business, by and large, seem to be more honest than Europeans. This impression reinforced an earlier experience in Bochum. We bought a second car, secondhand but with air-conditioning, a Benz, a 1982 380E, to take back to Durham at the end of our year there. We bought it but then left it for several weeks at the dealership, large, gleaming (even the workshop) and apparently respectable, waiting until we could drive it to a port in Holland to be shipped home. Only when we picked it up in the U.S. did I notice that the tires looked much more worn than they were when I bought the car. I think the dealer swapped the original tires for older ones when the car was sitting on his lot. Of course, if a tire had failed while we were driving to Rotterdam at high speed on the autobahn, the results would not have been good.

We did have a wonderful holiday starting from Konstanz. We drove south through the Alps, and then via the Gotthard Tunnel, to northern Italy. The tunnel is long and Cindy was a bit nervous, but we were soothed by the music of Peter Gabriel's hypnotic *Steam*, which I will always associate with that trip. The Italian lakes are stunning but we gave Como and Maggiore a miss and chose to stay at one of the smaller ones, Lake Orta. I later found out that my Duke colleague Irving Diamond, always one to find the very best, had stayed in an elegant hotel that had been converted from a monastery. We stayed in a B & B where the room was barely bigger than the double bed. On the other hand, the bed faced a big window that looked out on to the lake and the Isola San Giulio in the middle of it. So, not bad.

Chapter 15: Mexico, Ribeirão Preto

In the early 1970s I got a letter from W. K. "Vern" Honig at Dalhousie University asking me if I would be willing to collaborate with him on a project. Vern had been a student of Norman Guttman's at Duke and now was a professor in Nova Scotia. I knew him from meetings and from a sabbatical year which I spent in Oxford and he in Cambridge. A high point for Vern, a talented viola player, was the opportunity to play in the glorious King's College Chapel. In 1966 he had published *Operant Behavior,* a collection of invited chapters by notables in the field. It was for a while the definitive sourcebook on the topic. Now he wanted to do a new edition and asked me if I would be co-editor. I was happy to agree, even though I knew it would be a lot of work and to some extent a distraction. It prompted me to write a chapter on so-called schedule-induced behavior that reviewed the evidence for behavioral competition in operant schedules.[1] The new book *Handbook of Operant Behavior* was published in 1977. It was a success (for an academic book) and sold at a steady rate for many years.

Alas, Fred Skinner didn't like the book at all. It "was a disappointment". He quotes a review by George Reynolds, a brilliant experimenter but also a good Skinnerian: "Many will relish the invasion of Skinner's box by the tired troika of encephalization, S-R connectionism, and species-specific behavior. But it is yet moot whether such theories of learning are really necessary." Skinner adds "… when the book was published I put my copy on the shelf unopened, and there it remained. Dick Herrnstein said, 'You must be immensely pleased', but I was not".[2] I'm afraid the "unopened" comment says more about Skinner than the book.

The book helped make my name well known to the international community of operant conditioning people, a community especially strong in Latin America. I met Arturo Bouzas, a Memorial Hall graduate a few years after me, at the *Second Harvard Symposium on Quantitative Analyses of Behavior*. We shared many interests and he was now a professor at the Autonomous University of Mexico (UNAM) in Mexico City. He invited me to teach some classes at the university and we all went there for a month in the Spring of 1981. I'm not sure how satisfactory the classes were but the trip was fascinating. Wonderful,

---

[1] Staddon, J. E. R. (1977). Schedule-induced behavior. In W. K. Honig & J. E. R. Staddon (Eds.), *Handbook of operant behavior*. Englewood Cliffs, N.J.: Prentice-Hall.

[2] Skinner, B. F. (1983) *A matter of consequences: Part Three of an autobiography. New York: Alfred Knopf.* p. 368.

welcoming people and a great chance to explore a little of Mexico. Jessica and Nick were old enough to appreciate the country – and its problems. We stayed in the apartment of Jorge Martinez and his wife, a faculty couple who were away on leave. Near the pretty neighborhood of Coyoacan, close to the behavior analysis laboratory, it faced north and was actually quite chilly even though it was, nominally, early summer. Mexico City is at an elevation of nearly 7400 feet, so quite cool, especially at night.

The biggest shock was that the faucets ceased to flow after a day or so. Unfortunately, this was routine. Water was stored in a tank on the roof – a standard arrangement in countries where the temperature rarely goes below freezing. What we didn't realize is that the water flowed – trickled, actually – into our tank from the municipal supply starting at about two in the morning – but for only an hour or so. We had been living on the accumulation while the owners were away. As soon as that was exhausted, we had to live within our means.

My hosts insisted that we not drive ourselves. But they very kindly drove us to a number of places, like the Aztec pyramids of Tenochtitlan, and close to the top of the 17,000 ft volcano Popocatépetl, about 40 miles from the city. Because of the thin air on the slope of Popo, I was exhausted after walking just a few feet. I was not breathless, just weak – because breathlessness is caused by the accumulation of carbon dioxide, not lack of oxygen. My host, of course, was OK, living and being acclimatized at 7400 ft. Less exhausting was the famous Anthropological Museum. My chief memory, after Montezuma's feather headdress – now an object of contention with Austria, apparently – is a picture of daughter Jessica acting much too mature for her 11 years.

Mexico City is an extraordinary place. The roads were always crowded. One reason was that office workers usually drove home twice a day, leaving before midday for siesta, then returning in late afternoon and working well into the evening. Stoplights offered suggestions rather than commands. Drivers tended to go straight through a red light at night, when they could see in advance by its lights whether a car was coming. They were more cautious in the daytime, but tended to trust what they could see of the traffic rather than lights and signs – a more sensible policy than ours, in some respects. Traffic jams were common. Drivers were stuck for so long that buskers often provided entertainment in the middle of each intersection. The most dramatic were the fire-eaters. Surrounded by stalled cars on four sides, they would fill their mouths with kerosene, spit it out, light it and then blow flames in all directions. Health and Safety? *Basta!*

My student and now colleague Alliston Reid arrived to teach a few months after we left Mexico. Alliston spoke Spanish and he taught at the university for several years acquiring in the process a lovely Mexican wife. Alliston and Leonor were married in his mother's house in Cheraw South Carolina – a town possibly most famous as the birthplace of jazz genius Dizzy Gillespie. The pastor was Alliston's undergraduate teacher, John Pilley. It was an extraordinary scene. Leonor, who looked like an Aztec princess dressed all in white, could speak little English. She was surrounded by old, or at least older, white South Carolinians who spoke no Spanish. But all went off beautifully in a most moving ceremony. Alliston eventually returned to the U.S., teaching first at Eastern Oregon and then back at his alma mater Wofford College in Spartanburg.

Pilley and Alliston, many years later, achieved fame through Pilley's border collie dog, Chaser. Inspired by an Austrian experiment which had shown surprising learning abilities in a border collie, they designed another study to answer a couple more questions. On command, Chaser learned to pick a named toy from a group of eight or so in another room. Pilley, who was retired and thus had the time as well as the patience needed, kept adding toys over years of training until Chaser showed that she had learned a thousand or more different names. She also learned to retrieve a toy in a category, and to match a new toy with a new name and to respect the difference between a noun and a verb. For example, "take lamb" or "take nose", where 'lamb'; and 'nose' are different objects. Asked to "fetch Darwin" when "Darwin" was a name she had never heard before, Chaser would pick out the only novel object in a group of eight. She could fetch a "toy" or "Frisbee" or "ball" from a collection that included several examples of each category. She did about as well as human infants. All this was really quite remarkable, and Chaser's exploits caught the attention of the media, from NOVA to CBS's popular *60 Minutes* show. With the aid of a co-author, John Pilley wrote a book about it all: *Chaser*.

In 1985 I got a letter from a gentleman called Assistant-Doctor Professor José Lino Oliveira Bueno to say that his application to CNPq, the Brazilian science agency, for funds to support a longish visit by me had been approved. Lino and I had met earlier at a meeting and discussed the possibility of a visit, but I thought the prospect was rather unlikely, given the high cost. So I was delighted to hear that we now had support. The itinerary was rather ambitious: Rio de Janeiro, São Paulo, Ribeirão Preto, Brazilia and Belém, on the Amazon River. This trip was the beginning of a long friendship with Lino and his wife Belmira.

So, on the first of May 1986 Cindy and I set off on her first international trip. We flew from Miami to Rio de Janeiro on Varig, the Brazilian national airline. That was just the beginning. I can still hear the announcements in Portuguese: "Varig-e Vol..." – "Varig Flight..." as we waited in the Rio airport for our next flight, to Ribeirão Preto, the branch of the University of São Paulo where Lino taught. Lino met us at the airport and accommodated us royally. We had a luxurious bedroom in his apartment – I realized only later that he and Belmira had given up their room for us. Their buxom maid Leonice also looked after us very well. (I heard a few years later, that young Leonice had had cosmetic surgery – breast reduction – and was to be married. Cosmetic surgery was common even then in Brazil at every level of society. No wonder those Brazilian women are so perfect!)

Top left: **Son Nick, early 1980s**

Top right: **Nick working in Russ Church's lab at Brown University in 1986 or so.**

Daughter Jessica, age 11, in the Anthropological Museum restaurant in Mexico City.

The Brazilian economy was in a strange state during our first visit. They had just passed through a period of quite high inflation and to correct it, the government had imposed price controls. I suggested to my Brazilian colleagues that this would just cause shortages, but, during our brief visit, everyone was optimistic. That changed a few years later. Brazil has been through no less than four currency devaluations between 1986 and 1994, each time valuing the new unit at 1000 times the old: cruzeiro, cruzado,

cruzado novo, then back to the real, the currency first introduced in 1690 or so.

BRAZIL and MEXICO: *Top*: **Lino Bueno, in São Paulo, Arturo Bouzas in Mexico City.** *Bottom left*: **JS and friends in Guadalajara (curated by Carlos Aparicio)** *Bottom right*: **César Ades and friends in Brussels for the International Ethological Conference.**

Skinnerian psychology was a dominant influence in Brazil, which is in some respects the "Australia of psychology." Everyone knows how in Australia marsupials, a mammalian order little represented elsewhere, have radiated into niches occupied in the rest of the world by placental mammals: a marsupial wolf, marsupial deer (kangaroos) a marsupial rat, and so on. Skinnerian psychology in Brazil has radiated into niches occupied elsewhere by other intellectual species. Thus, in Brazil we can find behavior-analytic cognitive psychology, behavior-analytic developmental psychology, behavior-analytic personality psychology, and so on. It is a sociological version of the evolutionists "founder effect".

Cognitivist critics of Skinner may take comfort in the Australian analogy, feeling that they represent placentals to Skinner's marsupials. But celebration is premature: the question of extinction is not yet settled!

One of the founders of behavior analysis in Brazil was Fred Keller (1899-1996). Keller and Skinner were fellow students at Harvard. Keller, a gentle and amiable man (although "he could be tough" I was told by one of his students) went on to Columbia University where he became a much-loved teacher. He and William "Nat" Schoenfeld, a younger but somewhat less lovable Columbia colleague, wrote the first Skinnerian text in 1950: *Principles of Psychology*. A fellow Skinnerian called Keller and Schoenfeld "– in some respects the Lenin and Trotsky of our movement to overturn the accepted order in psychology". Keller visited Brazil for a year in 1961 (the first of several such visits), learned Portuguese and helped found the Department of Psychology in Brazilia in 1964. The Skinnerian approach was carried forward by Carolina Bori and (until a military takeover a few years later which devastated the universities and destroyed the lives of many academics) by Rodolfo Azzi. Skinner and Keller were not the only influences on Brazilian psychology, of course. But the role of radical behaviorism as the founding approach has produced interesting intellectual conjunctions – between Skinnerian and Piagetian approaches, for example.

Keller retired with his wife Frances to Chapel Hill and we saw them quite frequently, usually with other visitors, such as Lino and Belmira and some of my students and, on several occasions, my parents when they visited us from England. I recall one meal, at a rather fancy local hotel, the *Europa*, which entertained guests with a pianist. Fred and my mother, both in their eighties, got along famously. At the end of the meal, the pianist was through. Fred, learning that my mother could play, urged her to do so. And she did, going to the piano and playing for the assembled diners one of her favorite 1920s-vintage songs (*Lullaby of the Leaves* was a favorite). A wonderfully memorable scene. I was reminded of it again recently when Alex Kacelnik sent me a link to a short movie starring his own 94-year-old mother. *El Ruido* recounts her funny (fictional) reaction to noise pollution in crowded Buenos Aires[3].

My experience with Fred Keller's co-author Nat Schoenfeld was equally memorable but less agreeable than these. Nat gave a talk – quite good, I thought – at UNC Chapel Hill. It must have been sometime in the early 1970s, not too long after we had published the 'superstition' paper and several follow-ups which involved observing rats and pigeons as well as recording their operant behavior. At the end of the talk, I asked in all innocence, "Did you watch what the animals were doing?" – because we

---

[3] http://elpais.com/elpais/2015/04/21/videos/1429634437_930959.html

had recently found such observation to be very informative. Nat obviously thought this a hostile question and responded aggressively, berating me at length for … something. He was very upset and defensive, which much surprised me since my question was not hostile at all. Once again, I was shocked by the sensitivity of the eminent.

I gave a series of eight lectures in Ribeirão Preto. Then we flew to São Paulo to visit the main campus. I gave a talk there and met many colleagues. One of the most memorable was César Ades. Born in Egypt but living in Brazil since the age of 15, César was man of wide ethological and psychological interests. He was particularly interested in arthropods – spiders, especially. Since I had a long-repressed interest in invertebrates and other 'primitive' creatures – the fewer cells the better – I had much to talk about with César.

In 1996, Cesar visited Duke. We had just moved to a house larger though less ecologically correct than Kent St., at 1535 Pinecrest Rd. Empty cardboard boxes were still scattered around the front door and rolled up rugs were in most of the rooms. Cesar was staying with us and I took him in the afternoon to Duke to give his talk. Then I brought him back to the house. As we came through the front door we were greeted by loud and rather disagreeable music and an agitated Cindy. Apparently a large snake had come into the house, presumably through the often-open front door, slithered down the carpeted half-flight of steps into the open dining room and terrified Cindy. It then departed for the downstairs bedroom, at which point Cindy shut the door, stuffed a towel under it, and put a loud radio next to it to discourage Mr. Snake from returning.

"Well" I thought "César works with nasty things like spiders. He should have no problem dealing with a snake." So we gingerly opened the bedroom door and, sure enough, there was the snake poking his head up between the two single beds and looking at us is if to say "You got a problem?" Pretty soon, he realized we were interested in him and he slid off, into a rolled-up carpet. Quick as a flash, intrepid César grabbed the roll and started up the stairs with it. About half way, his voice went up an octave: "He is coming out!" he squeaked. And, sure enough, the snake's head began to emerge from the top of the roll. But by then we were all outside and let the snake wriggle away – into an empty cardboard box. I quickly closed it up. We removed the box, snake and all, and took it in the car to an undisclosed location.

The snake, though large and sinister-looking, turned out to be a harmless Black Snake – a friend, actually since he fed mainly on small rodents and competed with poisonous copperheads. We later found one other resident snake during that period, although smaller and less threatening-looking, living behind the curtains in the drawing room. We soon acquired an outdoor blacksnake who lived behind the house; we called him "Fred" and left him alone to pursue his good work.

Copperheads were pretty common in our neighborhood. I remember one occasion when I drove down our driveway and heard a slight crunch as I went. I assumed it was some kind of nut that I'd driven over. When I returned I saw that in fact I had run over a well-camouflaged copperhead. The crunch was caused by his distended belly, which contained his rodent lunch. On an earlier occasion during a morning walk near Lake Michie, a local park, I saw four snakes: two rattlers and two water snakes. Snakes are very common in NC.

Those were the days before flat video screens. My computer monitor was a bulky, warm, CRT screen and it sat in on a large shelf in the corner of my study. You couldn't see behind it without some trouble. When I had occasion to move it weeks after the snake incident, I noticed on the shelf detritus clearly recognizable as snake poop. It appeared that Mr. Snake had been resident in the house for much longer than we thought and had found a nice warm sleeping place. Moral: Don't leave your front door open in North Carolina in the summer, especially if it is at grade level.

Tragically César Ades died in 2012, at the too-young age of 69, struck by a car near his home in SP, a great loss.

João Claudio and Silvia Todorov in Brazilia.

The next leg of our Brazilian trip was to Brasilia, that extraordinary capital built hundreds of miles from the nearest big city in the middle of the jungle. The city – streets and public buildings – was built in just over three years and was ready in 1960. The chief architect was Oscar Niemeyer, a native Brazilian who adopted the name of his German grandmother. Brasilia has grown to a city of nearly three million people. Apparently Niemeyer – a Marxist cigarette smoker who lived to almost 105 – didn't like straight lines or ornamentation. He worked with marble, but the military dictatorship that overthrew Niemeyer's sponsor, dictator Kubitschek, preferred concrete and removed the marble from some of his buildings. Niemeyer's colleague Lúcio Costa, who laid out the street plan, liked to put like with like. When we were there, we noticed all the hotels seemed to be in the same area, for example, a senseless arrangement. Brasilia is not respected for its town planning. It is admired for its fluid modernist architecture. The National

Congress Building consisted of two bowls one right side up (the Chamber of Deputies) the other upside down (the Senate). We were free to wander over these fascinating buildings unrestrained by safety railings. The great Cathedral, which was full of evangelical worshippers when we walked in, is entered by an underground tunnel that ends in the middle of the building. This is so that its symmetry is not spoiled by an entrance. In Brasilia, all is form – magnificent in many cases, to be sure – over function.

Our host was João Claudio Todorov who got his PhD at Arizona State, one of the few U.S. universities specializing in Skinnerian methods. João Claudio had achieved some eminence in the University of Brasilia. He was a charming and rather charismatic figure, slim, usually smoking a cigarette and with the kind of deep voice I will always associate with Latin American soap-opera heroes. He and his wife were most hospitable and we were sorry to leave Brasilia so soon.

Our next stop was Belém, at the mouth of the Amazon River. The plane from Brasilia was small and we landed on at least one dirt airstrip, lined with fuel drums, before reaching Belém. It reminded me of Fort Rosebery and its Dakotas. The passengers were an interesting mix with a wild-West flavor. There were several men with boots, mustaches and cowboy hats. Most noticeable was an excited group of teenage girls twittering like magpies. They scampered through the cabin chatting up other passengers. They were especially interested in a priest, dressed in a long gray robe and a monk's cowl. Judging by the girls' facial expressions, he emanated not so much the odor of sanctity as just…odor. Eventually a couple of the girls' approached us. One admired my red pinky ring – a gift from a girlfriend many years ago. They spoke almost no English and we spoke about the same amount of Portuguese. But they did establish that we came from the U.S. Eventually one of them, Gisele, managed to ask "Do American boys like *mulattas*?" This earnest question, tinged with racial anxiety, from a sweet, young almost-child, touched me strangely. We assured her that they certainly would. It turned out that these kids were a volleyball team from Macapá in the North, on their way home after a match in Brasilia.

At Belém airport we were met by two people from the Psychology Department. One was a Central European called Ivo, the other a ruddy-complexioned American, Lee. They were like something out of a Conrad novel or *Apocalypse Now* – Europeans who had 'gone native'. Lee was unmarried but announced with some pride that he had recovered from venereal disease some five times. Ivo had a child bride who they termed 'Ivo's last chance' – to have his own child. They took us to a hotel and I got ready to give my talk that evening. The hotel was OK, but every slightly sticky surface in that damp tropical climate felt as if it was home to a thriving colony of micro-organisms.

Night falls swiftly in the tropics. By the time I was to give my talk, it was dark. We were in a large lecture hall, brightly lit by fluorescent tubes. The many windows were wide open with the surrounding jungle just a few yards away. There were no fly screens, yet, amazingly, there were few bugs and no mosquitoes. I wondered if the university doused everything with insecticide, but, apparently not. My Duke colleague Knut Schmidt-Nielsen was on an expedition a thousand miles higher up the Amazon than Belém, near Manaus, around this time. He also reported an absence of mosquitoes, although he did suffer from things called pium flies, which are smaller but equally blood-sucking. But not in Belém, at least during our trip.

BRAZIL 1987 *Top*: **The Macapá volleyball girls (Gisele on the left), on the plane to Belém.**

Bottom left: **Jose Carlos greeting the jealous woolly monkey (*right*).**

We met the extraordinary chairman of the University of Pará Psychology Department, José Carlos. He had moved to Belém from UNAM in Mexico a year or two earlier and, apparently, brought with him a bunch of computers that the Belém students could use. It was not clear that he had permission from UNAM for this transfer. José Carlos had some interesting pets. One was a sloth, damp, warm and immobile, all features that allowed the growth of green algae over much of its body. Another was a large Woolly Monkey. As we both stood in front of his cage, José said "He is very jealous" and illustrated his jealousy by putting his arm around my shoulder. The monkey instantly sprang screaming to the front of the cage trying to attack me. Nice! José Carlos also had a wife, Marilice. She looked more like a working girl than a faculty wife and was apparently notorious in modest Mexico City for walking the streets rather skimpily attired.

BRAZIL *Top*: **Cindy and friends in the Belém market.** *Below left*: **Wonderful singer Mariela Maça.** *Below right*: **Maceió beach and barracas.**

José Carlos was a talented fellow and described a number of interesting experiments he and his students had done. But then he described why he did not intend to publish them. This was a considered policy on his part, following a philosophy that was incomprehensible to me. What is the point of an experiment if the result does not add to the common store? Is science not cumulative? Was José Carlos anticipating the rise of contemporary narcissism? Is science to become just another form of self-development? Maybe he just didn't like criticism from reviewers.

The next day, we explored the city. Belém is at the mouth of the Amazon, which stretches to the horizon. The riverfront is more like an ocean front. All manner of animals were for sale in the market, monkeys, exotic birds, rodent-like creatures (Capybaras?) as well as magic charms – *macumba* – of all sorts. The covered market, made of cast iron – in Glasgow, I noticed with amusement – had fish of every conceivable type, some very large. We left for Rio the next day, from Belém airport, not air-conditioned and open on every side.

Rio is of course incredibly beautiful; we wandered along Copacabana beach with its wavy tile sidewalk. We admired the human scenery. We were harassed by a tame macaw on the way to the ride up the mountain.

We returned several times to Brazil. Odd events stick in memory. On one occasion, for example, Lino was driving us from the airport in SP to RP, a four-hour drive, when I realized that I had left the power supply for my (then very clunky) computer at the airport. The computer was necessary for our collaboration, so getting the power supply quickly was a must. I thought it might be on the wrong side of the security apparatus. Finally Lino got his brother, a person of some influence in SP, to go to the airport, find the thing and send it on to us in RP. On another occasion, again driving from SP airport, we went by mistake through a poor and heavily polluted industrial neighborhood. Cindy noticed that during our brief transit – no more than 30 minutes, surely – her silver wedding ring had completely corroded. On another occasion, Lino drove us at night to his mother's house in Araraquara about 60 miles from RP. With no habitation for much of the drive it was absolutely pitch black. The night was crystal clear. I have never seen the stars, the Milky Way, so clearly painted on the sky.

On an early visit, Lino invited us to his country club in RP – these clubs are an important feature of middle-class life in Brazil, which at that time resembled colonial Britain in some respects. Like English clubs in India or Africa they have tennis courts (very many in this case) a swimming pool and, of course, a bar. They host dances: Cindy and I danced to great Brazilian music as I puffed on a splendid local cigar. Smoking would not now be permitted, I think ('elf and safety' rules even in the hedonistic southern hemisphere.)

As part of one trip we were able to take a week's holiday in the Northeast, at a coastal town called Maceió. We stayed at a nice hotel, recommended by Lino. No one spoke much English, but we had a great time. We chatted to a couple who had come from Salvador, another coastal town not far away. Why had they come to Maceió rather than just going to the beach at home? Crime – Salvador beaches were not safe, apparently. On our beach were little thatched *barracas* – bars and cafes, many with entertainment. One singer especially, Mariela Maça (she gave me her card), was absolutely wonderful. I have often wondered how she got on.

This was a time of very high inflation. I noticed that no one ever had change – people kept as little cash as possible because it devalued so quickly. The banks, mostly large and ornate, were usually crowded with people depositing or withdrawing from high-interest accounts that partially compensated for the inflation. We needed Brazilian currency but the official exchange rate overvalued the cruzado many times. Language was not a bar to contacting a black-market guy who arrived at the hotel on a motorcycle and exchanged our dollars for a more realistic number of cruzados novos.

Chapter 16: Shiraishi

My college friend Phoebe is both venturesome and compassionate. She spent some years in Nepal as a VSO ("Voluntary Service Overseas) teaching English ("Shakespeare and knitting"). During her time there, her roommate died of amoebic dysentery. She later moved to Japan with her son and taught in Kyoto, under slightly more comfortable conditions. More recently she did volunteer work in the mental health wards of a dank and under-equipped Romanian hospital. I was invited to give a bunch of lectures and talks in Japan while she was there and at the end of that trip Cindy and I met up with her for a short holiday.

I had been invited to give a talk at Kyoto University and to chair a session at the International Ethological Congress in August of 1991. My friend and former thesis adviser Dick Herrnstein was to be in Kyoto at that time – to give a talk and also because one of his sons was to wed a Japanese girl. Cindy and I had to find a hotel in Kyoto. I consulted Mark Konishi a Japanese-born Dahlem friend, and he said "Stay in a Western-style hotel, not a *ryokan*". *Ryokan*, traditional Japanese inns, usually involve sleeping on mattresses on the floor and communal areas to encourage chatting among the guests. The food, of course, is also traditional. But Konishi recommended the Kyoto Brighton hotel, which was indeed splendid. Not typically 'Western' though. A pretty girl dressed like something out of a Great Gatsby movie, complete with pill-box hat and immaculate white uniform, was always there at the elevator – to press the button. It was embarrassing to see her perfection as we came in all hot and sweaty from walking around Kyoto in the August heat.

The hotel room had a number of electronic gadgets. Some of them had functions I could not immediately discern. In stores and other large enterprises, the public was usually addressed by a high-pitched female announcer over the PA system ("like talking ants" we thought) telling the assembled customers…something. Japan showed some fascinating, and puzzling, contrasts. For example, we visited a beautiful temple garden. Every detail was carefully chosen to be an elegant complement to the whole. Yet just outside would be an ugly tangle of electric wires, trash cans and other functional but unbeautiful stuff. In contrast, Western historic sites – a Cotswold village, a colonial Williamsburg, or even a Disney Epcot – meticulously conceal any intrusions of modernity.

We were accosted in the street by a small group of Japanese schoolgirls, anxious to try out their English. Was this an assignment, I wonder? Or were they just showing initiative – I didn't think to ask. This was outside the Nijō Castle, a 17th century Kyoto tourist landmark. Most memorable feature: wooden floors designed to squeak, to warn of potential attackers or unauthorized visitors to the palace concubines. This is the first time I heard of squeaky floors as a feature not a bug.

JAPAN Dinner in Kyoto in a riverside restaurant, 1991: JS and Dick Herrnstein with Japanese hosts.

Masaya and Naoko Sato with JS and Cindy in front of Prof. S's house in Tokyo. Plaque behind commemorates Prof. Sato's father, famed writer Haruo Sato.

Dick H stayed in a *ryokan*. I found out later that it was not a happy choice; Konishi was right: pick a Western-style hotel. Dick and I lectured back-to-back one time. I still remember the introductions: *Sensei* Herrnstein and *Sensei* Staddon – nice to get a little respect! We were also taken to dinner in a splendid restaurant overlooking the river. The waitresses were dressed in traditional kimonos and were almost embarrassingly deferential. After a while we noticed that Cindy seemed to be the only female guest in the restaurant…The food was not really to our taste, though obviously high quality by Japanese standards. A salty soup thing reminded me of bait water (the water you keep live bait in when sea fishing). This was our experience throughout our trip. *Soba*, a sort of noodle soup was OK, but by the end we were happy to seek out local MOS BURGER fast food outlets (a Japanese McDonald's equivalent). Our taste was definitely not Japanese.

After the academic events in Kyoto, we visited Phoebe and went with her for a couple of days to something called the International Villa on

Shiraishi Island in the Inland Sea. The Villa was part of an unusual enterprise by Okayama prefecture, to encourage foreign tourists: a series of traditional style villas, in beautiful areas, where locals can stay only when accompanied by a foreign guest. We took a three-hour train from Kyoto to Kasaoka. We just made it in a cab to the Shiraishi ferry. The boat, still hot even over water, glided through a fog, past small mountainous islands. The trees that wreathed the shoreline were filled with white egrets, like old-tyme Christmas-tree candles. The passengers getting off at Shiraishi were young – children and grandchildren coming from the cities, possibly a few vacationers, Phoebe said. Most of the residents on the islands were elderly.

We walked through narrow streets lined with old houses of stone or wattle, and newer ones of weathered wood. Every house had its raised garden of flowers and vegetables, and vines of bright orange trumpet-flowers and bougainvillea.

Just over a rise, not ten yards from where the walkway narrowed to a footpath, we were suddenly at the Villa. It was a fine modern one-storey building, with walls that looked like sliding paper panels. Inside were five bedrooms: four 'Western' style and one traditional. Phoebe took the traditional room we one of the others. There was but one bathroom and a single kitchen. We had a very pleasant two days on this lovely island, although there was one amusingly discordant note. We could smell a slight odor in the kitchen. When I researched the issue I noticed that the sink drainpipe went straight into the sewer. It lacked the S-bend trap that is universal in U.S. and U.K. plumbing and serves to block sewer gases from entering the room. I never found out if this feature was just a Shiraishi problem or is/was characteristic of all Japanese plumbing.

After Shiraishi we took a bullet train to Tokyo where I was to lecture for several days at Keio University. No more relaxing! We were guests of Professor Masaya Sato and were to stay at his house – an unusual honor for a foreigner in Japan, apparently. Masaya had pioneered the Skinnerian approach in Japan and was a major figure in Japanese psychology. His father Haruo Sato, a writer, had been a Japanese 'National Treasure' and it was so announced on a large stainless-steel board outside the block of four apartments that Sato owned. He and his delightful wife – a one-time student and now also a teacher, Naoko Sugiyama – lived in one of the apartments. This was a time of Japanese property boom: the Imperial Palace grounds in Tokyo were reportedly then worth more than all the land in California. Opposite Sato's residence was the home of Japan's largest builder, complete with parked Ferrari; it was a rich area.

Haruo Sato's original tea house had been taken apart and was reassembled in the basement of Masaya's apartment. As special guests, we were to sleep in that very tea house. As an honor, it left some things to

269

be desired, as we were to see. But we also fell short: we received several small gifts but, not knowing the custom, gave none. Japanese cities are full of gift shops designed to meet the need for gifts to suit every occasion.

Courtesy and consensus are big things in Japan. We were usually met and escorted from place to place, and rarely by just one person – usually it would be two or even four faculty who would meet us. It seemed clear that group action was the norm.

We entered his house after Prof. Sato. He removed his shoes and put on slippers; Naoko carefully turned his shoes around, to be ready for her husband's next departure, then escorted him to the sitting room, waiting until he was comfortable before returning to us. We likewise took our shoes off and got into slippers – which were different, we found later, from the slippers to be used in the bathroom. Professor Sato had everything very well organized. He seemed to have a checklist of things we were to do and see. High on the list was viewing a video of his elaborate wedding to Naoko.

I traveled to Keio University every day on the underground to give my lectures. I had to change trains, which was quite a challenge as I could read no signs – though after a while I got to recognize the character for the change station. This is one thing that made Japan so difficult to negotiate: without some ability to translate Japanese characters – *kanji* –the visitor is blind. In each station somewhere on the wall would be small name sign in alphabetic letters, so the task was not impossible, though it could be tough if the train was crowded and the view obscured.

My first impulse was not to agree to lecture in Keio. The reason was advice from another Japanese colleague, Arata Kubota. He was a talented and independent-minded chap with a real sense of humor – not a widely distributed trait in Japan, at least in my experience. I still remember a presentation he gave on a promising computational learning model – a 'Two-Ring Machine' (Turing machine). He thought I should say "no", possibly because he thought the experience would be too stressful. But after a while, perhaps responding to pressure, he changed his mind. He was right the first time, I'm afraid.

I lectured a small group of graduate students for three days – in English, of course. While the students could read English, their ability to translate spoken English was limited. A young Japanese colleague who had spent time in America, Takayuki Sakagami, therefore provided translation. I had to speak just a sentence or two at a time, with translation intervening. Nevertheless, all proceeded smoothly until the final day, when the students had to make presentations in English. The students found this difficult, but they all managed to complete the task – until the last one. He was an eccentric-looking young chap, he may have had colored hair – I'm not sure, but it would have fit his image. He did wear a sort of suede leather waistcoat that seemed to be on inside out – perhaps it was the fashion then, but it looked very odd. He had great difficulty in proceeding. I told him not to worry "Japanese is fine" I said, "Prof. S can translate". But it was obviously a matter of honor to him to do the job properly. Alas, he could not. By the end, he really looked as if he was ready to commit *seppuku*. Happily a fatality was avoided, but it was a difficult moment.

Our last day in Tokyo was for sightseeing. We were escorted by a very kind young faculty member, Toshio Asano. I particularly remember, in the tall Shinjuku district, looking down on a skyscraper with a glass-roofed swimming pool and a lone swimmer, and also 'Akihabara Electric Town' with street after street of electronic shops. We wanted to visit the Imperial Palace but alas got there too late.

At the end of a long day, we returned to Prof. Sato's house hot and exhausted – only to find that he had planned a farewell party for that very evening. Young faculty from many miles away were expected to come and, worse, the party was to be held in the basement of the house. Worse, because one wall of the basement opened directly into his father's re-assembled tea house – our bedroom. Escape was impossible!

The next day, only partially recovered from this excess of hospitality, we were escorted by Prof. Sato, Naoko and another faculty member to the Tokyo station where we caught a train for the almost two-hour ride to Narita Airport and the flight home. Whew!

The Japan trip was stressful but fascinating. I have never been in a country that felt so strange – not India, not Sri Lanka, not Brazil or Hungary. It wasn't just the *kanji*. The culture was very hard to figure out – perhaps that's not fair, after all we were only there for three weeks or so. But that is my impression. I think it is because of the social rules, their number, their strict enforcement and their remoteness from our rather slack and informal set of U.S./British rules. And of course the fact that we simply did not understand most of them.

Phoebe commented years later: "The key to understanding the Japanese is knowing that no one ever wants to be 'different' or stand out; so students won't answer a question until they are sure that everyone else has the same answer; no one will ever say "no", as this might offend. This can lead to confusion with foreigners in negotiation; "we'll consider it" means "no!" And I learned to watch the eyes in lessons; no one will admit to not understanding, so the glazing told me when I had lost them!"

Chapter 17: Administration

Some years ago, the scientific community was galvanized by the discovery of a new heavy element: Administratium *has 1 neutron, 12 assistant neutrons, 75 deputy neutrons and 111 assistant deputy neutrons, giving it an atomic mass of 312, all held together by a exchanging morons.* Administratium *is chemically inert, but has an anti-catalytic function. It tends to concentrate in large organizations, gradually causing paralysis accompanied by increasing demands for additional resources.*

In most medical schools, being Chairman (it usually is a man) of a department is a big deal. The appointment is typically for life and his authority is unquestioned. The Chair of Surgery is typically second in influence only to the Head of the medical center. It is quite otherwise on the 'campus' side. Academic chairs are often elected and usually serve for just a few years. Eddie Newman was the Chairman at Harvard when I was there, but Stevens and Skinner were the top dogs. An academic chair's role is the opposite of the harlot's: responsibility without, or with little, power. Consequently, academic departments often have difficulty replacing a chair. Every internal candidate is either uninterested or someone's sworn enemy. So, under the cover of objectivity, an outside candidate is sought.

Thus it was not because of my reputation for administrative genius that I have been approached for several chairmanships during my career. The geographical range was wide, from Binghamton in upstate New York to Leiden in the Netherlands and Perth in Western Australia.

I visited SUNY[1] Binghamton on a cold winter's day. Binghamton is not easy to get to. It's a small town in upstate New York, to the left of Albany and south-east of Ithaca. I was recovering from a collarbone I had dislocated in two (out of a possible three) places, tripping while playing tennis. (It's tough to get seriously injured playing tennis but, such is my athletic skill, I managed it.) So my left arm was in a sling and I was wearing shoes I had just bought that required no lacing. Nevertheless, my interrogation, though informal and amiable, was rather trying. I was meeting people all day and at a party later, until 11:30 at night. But the physical facilities, especially for animal research were very good.

I had an acquaintance at Binghamton, Norman "Skip" Spear, an animal developmental psychologist (why didn't Skip want to be Chair, I wondered?). We had both been on leave in Oxford during the 1973-4 academic year. He and I occasionally played tennis (Skip never sustained

---

[1] State University of New York

272

an injury, as far as I know.) He then moved on to Binghamton, attracted by excellent animal-research facilities and a way of life easier in many ways than his previous home at Rutgers in New Jersey.

The people were possibly more agreeable also. I have never been to Rutgers but, over the years, I have given talks at Princeton University nearby. One talk coincided with a peak in the popularity of a TV crime series located in New Jersey. Seeking to lighten things up, I began my rather mathematical spiel as follows: "Thank you so much for inviting me to Princeton. I am proud indeed to be here in this place just a few short miles away from the home of the great Tony Soprano…" This attempt at humor did not go over well with the Princetonians, who seemed to regard it as lèse-majesté. The Binghamtonians were a great deal less pompous but, despite the facilities and the good salary I was offered, I declined. I sensed that there might be problems to be solved in the Department – and, indubitably, the climate was not agreeable.

Sometime in the late 1970s I was considered for a professorship at the University of Western Australia in Perth. The experience of traveling to Perth for an interview was fascinating. On a long-haul trip the airlines in those days would put you up in a hotel overnight, in between connecting flights. I therefore arranged that on my trip to Perth I could stay overnight at Singapore on the way out and Hong Kong on the way back. On one of the return legs Cathay Pacific bumped me to first class. As part of some kind of company celebration, all the first-class passengers were given a little silk sachet containing a pearl – which I gave to Jessica, then aged perhaps eight, on my return.

I was in Perth for about a week. The chairman, Aubrey Yates, a student of larger-than-life English psychologist Hans J. Eysenck, was a welcoming host. The senior figure at UWA was John Ross a Cambridge-trained expert in the visual system. But he was away on leave that year; I was very sorry not to meet him. Robin Winkler, now sadly deceased at a young age, was a talented clinical psychologist with Skinnerian inclinations but a broad perspective. Peter Livesey, a physiological psychologist who later spent time at Duke with my colleague Irving Diamond, was another notable researcher. The faculty in general were very hospitable. The memory that stays with me is of a little book, *Dinkum Dunnies* given to me by Judith Kearins. A *dunny* is Australian for an outside loo and the book is a collection of photographs of dunnies from all across…not the fruited plain but the eucalypted outback. Judith studied visual memory in aborigines so I daresay the book reflected her rural experience. It was a "bonzer" gesture and typified the good feeling I got from the trip.

A week or two after I got back to Durham, I was offered the job. I looked in some detail into the complexities involved in emigrating. I still have a file folder full of helpful Oz government brochures from that time. After some discussion, DB was invited so she could see WA for herself. But we could not accept that offer unless I was pretty committed myself which, in the end I was not. After much agonizing, I turned down the job and decided to stay at Duke where I was, in truth, pretty happy.

The air trip was much pleasanter than it might be today: minimal security, more space, better service, smoking. And the destinations... Singapore and Hong Kong were a fascinating contrast. Singapore then still retained much of its British colonial atmosphere. Red-tile-roofed low-rise buildings predominated, with just a few skyscrapers beginning to intrude. The island was under the benevolent dictatorship of Lee Kwan Yew, who rescued the state from its original home as part of Malaysia (Singapore is still dependent on Malaysia for its water supply). The story of Singapore as a separate state is extraordinary. After independence from Britain in 1963, the new nation of Malaysia soon came to regard the island as a liability. Singapore is majority-Chinese; the rest of Malaysia is Malay and other groups, many of them Moslem. After independence, ethnic tensions increased. Finally in 1965 Singapore, regarded as both a social and an economic liability by the Malaysian government, was expelled from Malaysia.

Singapore has no natural resources except, perhaps, its physical location. Lee Kwan Yew tried right up until the end to keep Singapore as part of Malaysia. Even he, the author of its subsequent huge prosperity, at that time saw little future for the tiny resourceless island as a separate state. But once it became independent, its relative racial homogeneity, its Chinese commercial traditions (many of the businesses in Burma during my Mother's time were Chinese who were, as was often said in those oh-so-insensitive times, "the Jews of the Orient"), its good location for sea-borne commerce, and the autocratic but not kleptocratic rule of Lee Kwan Yew, within a few years raised it to a height of prosperity unequalled in South East Asia. Singapore was about halfway there when I arrived in 1977 or so. Now (2014) Singapore is ranked 9th in the world in GDP per capita, just ahead of the U.S. Are there lessons here for the Brexit[2] debate?

I stayed in Hong Kong on the way back from Perth. Arrival was quite dramatic. I woke up as the plane was approaching Kai Tak airport, slightly alarmed to see hills *above us* on both sides. Landing at Kai Tak was apparently quite tricky. It was closed in 1998 and Hong Kong is now served by two other, larger airports. Hong Kong at that time was still under British rule. It was known for finance, cheap knock-offs of mass-market products and tax-free shopping. I remember the grand Star mall –

---

[2] The continuing debate over Britain's place in, or exit from, the European Union. http://www.ft.com/intl/cms/s/0/16787d98-7293-11e5-bdb1-e6e4767162cc.html

now dwarfed by many others – where I bought a couple of quartz watches (very high-tech for the day) for DB and myself. DB didn't like hers (a Cartier-type lady's watch) and preferred mine. I traded hers for another for myself at an obliging Raleigh store – they made a small profit on the swap.

Hong Kong's 'built environment' was striking. Narrow skyscrapers soared above tiny plots of land. The mountainous surroundings are dramatic. There were at that time just two universities in Hong Kong, one taught in English the other in Chinese. I briefly visited the English one, high on the side of a hill, and chatted with Geoff Blowers, a British psychologist who still works in Hong Kong, now at Hong Kong Shue Yan University. Apparently the Department at Hong Kong University would have been happy for me to give a talk, had I let them know in advance. An opportunity missed, as I would have liked to stay a little longer.

I revisited Singapore in 2010, partly to see the new Duke-NUS[3] Graduate Medical School. Most of the colonial traces had vanished, except for a few tourist attractions like the Raffles Hotel. The Raffles, named for Sir Stamford Raffles (1781-1826) founder of the colony, was massively restored in 1991. The famous Long Bar, which at one time hosted writers and celebrities from Ernest Hemingway to Rudyard Kipling and Somerset Maugham had been physically moved to a new location. But the city is now dominated by massive new structures, especially a huge hotel that consists of a line of three tall buildings with what looks like the Eurostar stretched across their tops. It was under construction when we were there in 2010.

I returned to UWA in 1998, some twenty years after my first visit. The occasion was a collaboration with Clive Wynne, who had been my postdoc when were both in Juan Delius' lab in Bochum in the mid-1980s and was now on the UWA faculty. Perth is a very pleasant modern city, surrounding the beautiful broad Swan River. The river, which is more like a lake in places, is largely free of commercial docklands because of its shallowness. It is a wonderful resource for the city, as are the indigenous black swans (the ever-creative Nasim Taleb would have found a different name for the rare events described in his famous book, had he spent time in Perth). We spent a few days traveling by rented car to the south, visiting wineries, the amazing tall Karri and Jarrah forests, the Valley of the Giants Treetop Walk and the like and the old whaling town of Albany. We explored the downtown area on the free circulating bus. Occasionally a sleeping and apparently intoxicated aborigine could be seen dozing in air-conditioned comfort on the back seat.

---

[3] National University of Singapore

The visit went well professionally, but the atmosphere in Psychology was very different this time. I recognized only one faculty member from my first visit. Doubtless there were others around, but they had either retired or I didn't see them. The one who remained was an alcoholic, or so I was told. He seemed perfectly sober to me, but was not I think happy with the current situation. Clive and a few other young faculty, mostly American, seemed to be rather isolated from the 'old guard' who ran the department. Clive is an energetic chap, a great addition to any department, I should have thought. But possibly his very energy and innovativeness rather threatened some of the older faculty. I don't know. But some things happened that hinted at real problems.

There is a faculty custom in English and Australian universities of meeting for coffee at 11:00 or so each morning. It's a nice habit, which I have observed in Oxford, York and UWA. But in Perth I did have a rather odd feeling about some of these gatherings. People were affable enough but on several occasions as I entered the room the faculty who were there did a fair imitation of a dog who had just pooped on the carpet, looking away rather shamefacedly. I had no idea what this meant. I found out much later after we returned to the U.S.

UWA had paid some of my expenses for this trip. During the three months or so I was there, Cindy and I took a short vacation in one of the closer spots to Perth, remote as it is from the rest of Australia: the Indonesian island of Bali. Apparently a rumor circulated that I had somehow inveigled UWA to pay for this (by no means lavish) holiday. Worse, we had apparently traveled *first class* from the U.S., again at UWA's expense. I wish! It was all nonsense, of course. But I am still amazed at the credulity, paranoia even, of a bunch of well-educated, middle-class white folk. Shades of Northern Rhodesia and the African-blood-drinking English! The superstition of these 'Europeans' is an antidote to racism only equaled by our later experience with the white underclass in the North of England. Stupidity has no color.

We returned to Perth a third time in 2000, this time accompanied by a woman friend of Cindy's from Durham. It was a chance for Anne M to see Australia with us. We traveled via London, which posed a small problem. A couple of weeks earlier, over the Christmas period, Cindy and I had visited my aged mother, living alone in Cove, near Farnborough. (My father had died several years earlier.) Now we had to decide: Since we were passing through Heathrow, should we visit my mother again, albeit just for a day? We decided it would be too upsetting for all concerned, since the visit would have to be short – and we had just seen her a couple of weeks earlier.

Our trip was delayed because of a snowstorm in Durham, a rare event for which RDU airport was not at all prepared. We left a couple of days later than intended. Well, finally, here we were in LHR seated in the plane as the jetway pulled away on the second leg of the trip. Next stop: Singapore. Suddenly there was an announcement: "There is a problem with one of the passengers…" "Who is this wretched person! Delaying us *again*, just as we are about to set off" I grumbled silently. Well, it turned out that the passenger was me. My mother had just died, of the after-effects of a fall about which I, of course, knew nothing. My sister, then living far away in North Yorkshire, at first assumed that I was in Australia, which I would have been without the weather delay. So she called Clive in Perth, who knew about our delay, since he had to meet us. "They are probably at London airport right now!" he said. The phone system then was not the thicket of automated obstacles most systems have since become. Judy was able to actually get through to the gate. She told them I was needed in England and why.

British Air's response was exemplary – amazing. They rolled back the ramp so we could disembark; offloaded our luggage; took Anne's passport (which was with ours) back to her so she could continue the flight; and finally gave us a car and driver to take us to my mother's house in Cove. So I was able to take care of most of the matters associated with my mother's funeral and her estate – duties which had fallen entirely to Judy when my father died a few years earlier. We wrote BA a very appreciative letter. The unfortunate part was that Anne had to be shown around Oz by strangers. We finally got to Perth after some weeks in England, but too late to be with Anne, unfortunately.

My mother died just before her 96th birthday. She was a remarkable woman who had a remarkable life. She never lost her marbles. In her last couple of years her attention wavered a bit. She would occasionally fail to remember things, not because her memory was failing – it wasn't, any more than is normal in the elderly – she didn't remember because occasionally she didn't attend to what was being said. She wasn't particularly active in her later years, although she would ask to be taken for a walk each day when we were with her. She would spend much of her time in her last decade reading, watching TV – and smoking. She was rather shy but very self-contained – the local supermarket ladies called her 'The Duchess'. But she was gracious and sweet natured and well-loved by her Cove neighbors, who were enormously helpful in her last years. She was very smart, especially about language, as her *Recollections* show. In another time, she would have gone on to a distinguished career, in academe, perhaps. She was very ambitious for her children. She was upset with my sister Judy because she didn't 'marry up' – to one of those well-spoken, public-school-educated DAs in Fort Rosebery, for example. Judy chose to marry Bob Rawlinson, an engineer college friend – and from the North! Judy wrote her an anguished letter trying to explain and justify her marriage, which was happy and lasted until Bob's too-early death in his 60s. She and Bob and their two children, Jonathan and Michael, traveled all over Africa following Bob's work as a mining engineer. Judy has recounted her early life in Africa in *Finding a Flame Lilly.*

I don't think my father ever had much doubt about Bob, and our mother eventually realized the wisdom of her daughter's choice.

My mother's ambitions for me were initially rather modest – a nice civil-service job in the Forest Service, perhaps? But she was relentless in getting me into the best possible schools. She was not happy about my marriage to DB, but she didn't show it. Her unhappiness may have been somewhat mitigated by DB's good looks and professor father. I only found out much later than DB was not very nice to my parents when they were in Durham. On the other hand, both my parents were at my second wedding and they loved Cindy, so all ended well in the matrimonial department.

I saw little of my father when I was growing up because he was usually away from home, in the army. But in the eight years after the war, from 1945 to 1953 when he went to Africa, he was at home. He was a funny, kind and gregarious man. He had great practical talents that he had picked up I know not where. He never went to any kind of engineering or trade school, as far as I know. But he was quite skilled at woodworking and construction, what the Brits call DIY[4]. I learned much practical gadget-fixing stuff from him. As I mentioned earlier, he had a talent for coining the memorable phrase. He and I didn't do the traditional father-son sporting things. For one, thing my father had no great interest in professional sports; *ich auch!* – although I did have a mild interest in QPR[5] and occasionally walked up Lisson Grove from school to see the cricket at Lords. But both my parents played tennis, and I often used to cycle up the Edgware Road to the tennis courts in Edgware that William Moss provided for its employees. My father finally retired after his years of service in Zambia to a semi-detached house in Cove with a nice large garden. He died, of prostate cancer, in 1995.

My third brush with administration was Leiden University in the Netherlands. Leiden is Holland's oldest and most distinguished university. Established in 1575 it lists on its fame wall philosophers from René Descartes to Baruch Spinoza and more than a dozen Nobel laureates. Niko Tinbergen was educated there and the university in 1990 had a splendid program in ethology. My own acquaintance with ethology was limited to my time in Oxford and a handful of papers on topics related to adaptive

---

[4] Do It Yourself.
[5] Queens Park Rangers football team.

behavior and evolution. I published a large and well-received book[6] in 1983 that attempted to place animal learning in a real biological context, and I suppose that was noticed. It was nevertheless a surprise, in March 1990, to get an invitation, from a Dr. Jacob L. Dubbeldam, Chairman of the Selection Committee, to apply for the Chair of Ethology at Leiden.

Cindy and I visited later that year. Leiden is a beautiful old town with many of those typical tall, narrow Dutch row houses bordering the canals. In one of them lived a faculty member who collected art prints – etchings, wood-cuts, lithographs, etc. – from Eastern Europe. We bought two, from Czechoslovakia, then still behind the Iron Curtain. I spent much time talking with a faculty couple who shared some of the problems of the Department. After our visit, I got an offer of the job – a great honor. But the conditions were rather onerous. The salary was less than I could expect to make at Duke. I would eventually have to learn Dutch – I didn't relish negotiating with administrators in Dutch. And my circumstances at Duke had changed. I got a five-year, once-renewable "K Award" a grant from NIMH[7] that freed me completely from teaching. It meant that I could spend all my time on research. So I declined.

But administration did finally track me down in 1985 when I agreed to chair the Duke Department. Psychology at Duke has gone through several identity crises. Beginning as plain-vanilla Psychology it split sometime in the 80s into *Psychology: Experimental* and *Psychology: Social*; then re-united as *Psychology and Brain Science* and finally it become *Psychology and Neuroscience*. All this a result of the continuing conflict between the 'soft' social, personality and clinical parts of psychology and the 'hard' experimental and biological wing. A truce seems now to have been signed: Problem solved by rebranding all the 'hard' bits as 'neuroscience' – not that some of *them* are so hard…

I was asked to be chair starting the year I was in fact going to be on sabbatical leave in Germany. A temporary substitute was found and I began as chair the year I returned. Motto: Watch out who you get as an administrative substitute. The choice that year was not a good one "He always sounds like the last person he talked to who can present any kind of appeal to his 'good nature'" wrote a colleague.

I returned to a turbulent two years of administration. My first duty was to try and conciliate various faculty who I knew were opposed to my chairmanship. What a failure! One chap, a senior clinical psychologist, as I began my "let bygone be bygones" spiel, retorted in threatening fashion "I am a vindictive man" indicating that bygones were not gone for him.

---

[6] *Adaptive Behavior and Learning*
[7] National Institute of Mental Health, the premier grant-giving body in psychology.

I experienced that level of overt hostility only one other time, a few years later, with another colleague working in my own research area but with a different theoretical approach. I had supported his hiring because I thought a different point of view would be stimulating – and the guy's research record looked very good (now, after some years' close acquaintance – my last lab was near his and my collaborators had many opportunities to examine his work – now it begins to look more than a little dodgy). He was hired and put up for tenure a few years later – but turned down by the university-wide committee. I and couple of colleagues wrote letters in his support and the decision was reversed. Much the same scenario recurred on his promotion to full professor – turned down at first, supported by me and couple of other senior people, then eventually promoted. The obvious inference is that the review committee in both cases got information from external reviewers (who are always consulted in these cases) that was not known to us.

I don't think the candidate knew of our interventions. I didn't tell him, but after the incident I will describe, another colleague did give him at least a hint. All the animal-lab people were moved to a new building a few years before I retired. Since we had similar interests and were now neighbors, I suggested to my colleague that we might have some joint lab meetings. It turned out that he was not as excited as I was by the prospect of competing points of view. I was greeted not with enthusiasm but with a barrage of unprovoked insults – an extraordinary outburst.

I have been surprised many times to discover that some people are a lot more self-centered and nasty than one might expect. I think this is the experience of many who are thrust into positions of some authority: they are often taken aback by the unsuspected pettiness of their colleagues.

One incident during my chairmanship, although shocking from one point of view, was also humorous. Searching for our electronics technician one day I was finally told by one of the staff "He's at Al's house fixing his lights." Al was a clinical psychologist. His research was on sleep, but in fact he hadn't done any for many years. His teaching was well received, but his main activity now was the production of *collages* – graphic art. He has since had many shows both in the U.S. and abroad; he is evidently someone of consequence in the collage world. But appropriating our tech guy was – in the words now much favored for wrongdoing – "not appropriate". So I called him in. "Al" I said, "You can't send our technician to your house to do lighting for your art." Al was unabashed. "But my work is the most important in the Department!" he said, stamping his feet (really – or metaphorically – memory doesn't distinguish). "Maybe", I said, already beginning to grin, "but it isn't psychology…" We worked it out amicably. But toddler-level behavior is surprisingly common in adults, I've found.

I quit the chair after two years. I realized I was just living off intellectual capital. It was almost impossible to think through problems and do real creative work. It is always possible to publish though. A friend and distinguished colleague who rose to considerable administrative heights managed to maintain a steady rate of publication throughout his career. But originality, not so much. I am very suspicious of senior administrators – there are many such in medical centers – who boast of running several research projects. I doubt that their contributions add more 'signal' than 'noise' to the body of science.

Chapter 18: 'Possum'

In 1963, Clark Kerr the respected president of the University of California system, had a really bad idea: not university but *multiversity*. He was responding to the rise of the so-called 'research university', which departs from the traditional university in very many ways. Kerr was reacting to a trend – indeed, sensing wind direction seems to be the defining characteristic of our more successful university administrators. The trend was away from any commitment to universal knowledge, universal values or indeed anything universal at all about the university. Instead of being home to a quasi-religious community with a single creed – *veritas* – Kerr saw the modern university as something like a coral reef, an accidental colocation of disparate organisms, mostly invertebrate, united only in their need for external sustenance[1]. A loose confederation of unequals, some devoted largely to teaching (the lower echelons), some to 'basic' research, some to political activism – for the environment, say, or polysexual equality, or social justice. Still others are contractors, to the Federal government or big business, who find useful the cachet of the university, and its convenient access to willing low-paid people of talent – students, postdocs adjuncts and others in the ranks of the untenured.

The fragmentation of *disciplines*, as each scholar seeks to find his own uninvadeable market niche, has also played into Kerr's vision. In every academic department – especially the sciences, but also now the humanities – status is defined externally: by grants acquired and prestigious publications, by citation rates and 'impact'. Faculty look to their discipline – disciplines increasingly subdivided and overspecialized by each researcher's need to find a niche small enough to dominate – rather than to their institution.

Grant-getting pressure has led to another coping strategy: the formation of groups. 'Collaboration', a feature of many 'idea' businesses like tech or advertising, is favored. Interdisciplinary projects provide some counter to overspecialization and some assurance of stability of support. And of course big projects make the grant-evaluation process cozier, as one large bureaucratic entity – NSF or NIMH – communes with another – a center for mental health and brain function, say, or a project on cancer and the environment. All these factors tend to weaken the allegiance of individual faculty to the home university or even the home department.

The trend has not progressed quite as far in places like Oxford and Cambridge where faculty are affiliated with colleges and eat and meet

---

[1] A biologist colleague chided me about this analogy. He thinks that coral reefs have much more coherence and interdependence than the modern university.

with a variety of colleagues there. But there is nothing comparable in the American university. Indeed, it is tough to get support even for a single faculty dining room.

ACADEMIC DRESS-UP

Left: JS and Knut Schmidt-Nielsen in a sartorial mano-a-mano after a Duke commencement. Knut always won (the hat).

Bottom: An attempt to catch up: JS and five others getting honorary degrees from the Université-Charles-de-Gaulle – Lille 3, in 2004. JS third from left, Portuguese writer and Nobel Prizewinner José Saramago is third from right. This was an especially great honor because I didn't know personally anyone in Lille.

Kerr's vision has prevailed. The root idea of the university as a community joined together in the pursuit and dissemination of truth has been largely supplanted by all sorts of pragmatic functions to do with public betterment, social mobility and national competitiveness.

In the early 1990s I was invited to edit something called the Duke *Faculty Newsletter*. This was a modest 8-12-page pullout, appearing once a month in the *Duke Dialogue*, a university-wide weekly newspaper. Its main function was to publish the verbatim minutes of each month's Academic Council faculty meetings. But some space was left over and for a year or two previously it had been filled with the gentle musings of the editor and his friends.

I'm not sure why I was asked to edit it, but I do know why I agreed. There was no remuneration and it was only because I had some time free for 'service' – because I was now teaching much less – that could take it on. But I had grown increasingly unhappy with the fragmentation of the university and what appeared to me to me a loss of its core purpose – the change from university to multiversity. Three things were especially disturbing: attacks on the whole idea of truth; the almost complete absence of real debate about important issues; and the mindless near-worship of racial and ethnic (but not intellectual) *diversity*.

So I accepted the job, writing in my first editorial "If it is to be something more than a record of the academic equivalent of bridal showers and lottery wins, the *Newsletter* must be a forum for those in the Duke community who value what is universal about the university: the search for truth, commitment to scholarly values, to rationality and to civil debate."

Editing the *Newsletter* got me more involved with faculty politics than I had ever intended. I wound up being asked to serve as secretary to ECAC, the Executive Committee of the Academic Council, an elected group with an impressive title, much contact with the senior administration of the university but, alas, little real power. I served in that capacity for ten years.

The major incident during that time was of course the nationally, nay, internationally, famous Duke Lacrosse Scandal of 2006[2]. It fit a popular 'narrative'. In episode 1, a poor black single mother – like Tom Wolfe's 'honor student' in *Bonfire of the Vanities* – needs money, and is hired as a stripper by a frat house of rich white (well, almost white: there was one black) guys. That would be Crystal Gail Mangum, a student at NC Central University, a local HBCU[3]. Then, in episode 2, she is horrifically raped by several men. In episode 3, all the rotten, spoiled frat guys deny their guilt.

This terrible crime caused an uproar. Agitators bayed for action. African-American activist groups marched on the campus. A group of 88 mostly humanist faculty published an advertisement in the Duke student newspaper two weeks after the story broke. The ad was made up largely of quotes from anonymous students. The preamble said "We are listening to our students. We're also listening to the Durham community, to Duke staff, and to each other. Regardless of the results of the police investigation, what is apparent everyday now is the anger and fear of many students who know themselves to be objects of racism and sexism, who see illuminated in this moment's extraordinary spotlight what they live with every day..." So it is clear that the ad did not explicitly presume that the lacrosse guys were guilty. It was equally clear that its authors had little doubt about it.

---

[2] http://www.economist.com/node/9804134
[3] Historically Black College and University

Following the proceedings on the Executive Committee during the lacrosse scandal was an unsettling experience. I listened to the President and other senior officers agonize over what was a completely uncertain issue. For a long time no one knew the truth of the matter. Yet the pressure to act was relentless. Apparently the lacrosse coach had not done what he had promised to do about the 'culture' of the team. The administration's response was to use the crisis as an excuse fire him – an attempt, I suppose, to placate the "do something" cries of the mob. The lacrosse season was suspended – all before the facts were known.

One of the worst voices was Houston Baker, a distinguished professor of English and African-American studies, who did a plausible imitation of Tom Wolfe's Reverend Bacon (in *Bonfire* ...). His inflammatory speeches bore no relation to the private man I knew as an agreeable and conscientious renter of a small house of ours – he rented while his own more luxurious structure was being built in a nice suburb. (We had several tenants over the years. The most memorable was African-American dancer at the American Dance Festival. He was a sweet man. He left behind in the house two things: a thong and a bong. Dance culture, I guess.)

Our "bong-and-thong" dancer renter in 2000 or so.

One of the authors of the famous 'Group of 88' ad was a professor of English who later became a friend. I had no contact with her during this time and I still cannot understand how she could act as she did. I can't comprehend the Jekyll and Hyde characteristics of many people when race is involved.

The whole thing was based on a lie, of course. The prosecutor, Durham District Attorney Mike Nifong, running for re-election in a heavily black constituency, had ignored exculpatory evidence and was eventually disbarred – a mild punishment considering the fact that had he proceeded, three young men might have had their lives ruined. But it cost Duke in both reputation and cash settlements with the three innocent frat boys. Just how much cash we are not allowed to know. But since the President consulted the Trustees and all were on board with every wrong action, no one took a fall.

After all the lies were exposed what happened to the liar, Crystal Gayle? Well, nothing at all. She was left free…to a few years later fatally wound her boyfriend, something which might have been avoided, had she been prosecuted.

My turn at the *Faculty Newsletter* was 15 years earlier than the lacrosse scandal. Some people still remember *FN*. Elderly faculty in other departments, when I meet them after years of absence, seem to recall just one thing: the "V-P Count". The V-P Count is a simplistic measure of one aspect of the university: the number of people in the administration with "vice-president" or "vice-provost" in their title. It's a proxy for the growth of administration in general. It began with an informal survey, using old phone books, of the number of people listed as "faculty" vs "staff" over the years. I further subdivided "staff" into central administrative staff vs staff located in academic departments. Lo and behold! The ratio of faculty to staff had grown smaller and smaller over the years, and the growth in the denominator was all in staff *outside* academic departments.

I remember, for example, that when I first came to Duke the legal department consisted of one person; now it numbers I think about 15, plus nine or so secretarial staff. The number of undergraduates has of course changed hardly at all over that period. The story is the same for other non-academic units. The V-P count was a crude measure of administrative bloat.

To illustrate this disturbing trend, on the first or second page of every issue of the *Newsletter*, in large letters, appeared the current number of V-Ps: the "**V-P Count**". The number increased pretty steadily from issue to issue, of course. Finally, I got data from the Duke Archives and published a graph in November, 1991, plotting the number of vice-presidents every year, beginning in 1959 and ending in 1991. The graph rose with only one small dip, from 3 in 1959 to 19 in 1991. I'm not sure what it is now – especially as there has been some creative re-naming.

This inflation of the periphery is not entirely Duke's fault, of course. It has happened at all other universities. And it has happened in the federal government, which increasingly controls even private universities. All but a handful of U.S. institutions of higher education are critically dependent on federal funds. The chance of losing these funds is sufficient to make administrators exquisitely sensitive even to 'suggestions' they get from the federal Department of Education (as in, "Nice little university you got there; wouldn't want to find Title IX violations..."). This process has allowed a steady flow of regulations, some reasonable, many not, but surely excessive in number and complexity, to wash over universities. To deal with this flood, colleges and universities must employ a growing number of compliance staff. And Parkinson's Law applies, of course, to all these institutions, universities and government departments alike[4].

But the main objective of the newsletter was to encourage discussion. I tried to get some debate on the strange view of science espoused by

---

[4] http://www.economist.com/node/14116121

some humanists by reprinting in an early *FN* issue the edited text of a talk[5], "Science and Technology Matter in Liberal Education" given by colleague Matt Cartmill, a distinguished biological anthropologist – anthropology is a field nicely balanced between the sciences and the humanities. Cartmill's point was that many eminent humanities scholars portray science as purely political, a quest for power by (for example) a patriarchal elite. Underlying all is the idea that there is no such thing as objective truth. The fact that this idea undermines itself seems to have eluded these guys. Having abandoned truth, they feel no need to dwell on the truth or falsity of their own ideas.

Quotes made the point: Eminent French scholar Jean Baudrillard, for example, ends a chapter with a slogan from Friedrich Nietzsche: "Down with all hypotheses that have allowed belief in a true world!"

To her very great credit – almost no other humanists/ postmodernists/ deconstructionists at Duke were willing to follow her lead – Barbara Herrnstein Smith, of whom we have already heard (in Chapter 7), responded to Cartmill's piece in a thoughtful article, also published in *FN*. Entitled *Postmodernizing the Two Cultures: A Reply to Matt Cartmill.* She questioned Cartmill's conclusion, calling Baudrillard "a provocative culture-critic" rather than a philosopher, but praising Donna Haraway, also cited by Cartmill, as "an original and perceptive cultural theorist". Amusingly. Haraway is a recipient of (among many other honors) the J. D. Bernal Award, for lifetime contributions to the field. (You may remember communist genius Bernal from Chapter 1.)

To which Cartmill responded with a few more quotes – from Haraway: "The detached eye of objective science is an ideological fiction, and a powerful one. But it is a fiction that hides – and is designed to hide – how the powerful discourses of the natural sciences really work." Though scientists seek to conceal the fact, says Haraway, science is really a kind of imaginative literature: "Biology as a way of knowing the world is kin to Romantic literature, with its discourse about organic form and function. Biology is the fiction appropriate to objects called organisms; biology fashions the facts 'discovered' from organic beings". And the goal of all this science-fictional activity is political: "Biology, and primatology, are inherently political discourses, whose chief objects of knowledge, such as organisms and ecosystems, are icons (condensations) of the whole of the history and politics of the culture that constructed

---

[5] Science and Technology Matter in Liberal Education, given on October 10, 1991 at a conference on "Science and Technology in Liberal Education" hosted by Duke under the auspices of the Association of Graduate Liberal Studies Programs.

them for contemplation and manipulation. The primate body itself is an intriguing kind of political discourse."

Well, probably no meeting of minds between Herrnstein Smith and Cartmill. But this battle has I think been more or less won, in the sense that statements such as "there is no objective reality" and "science is basically a political activity" are no longer accepted without question, even in English departments.

I am impressed – and depressed – at a distance of 25 years, by how many of the topics raised in *FN* are still with us today. Take *sensitivity* for example, which is often used to justify censorship – censorship which is really caused by fear. In many cases, *sensitivity* is a cloak for *cowardice*.

In London recently, visitors to an exhibition celebrating freedom of expression found plenty of vulgar 'transgressions' but looked in vain for one work, listed in the catalogue, by an artist known as Mimsy. *Isis Threaten Sylvania* is a series of seven satirical light-box tableaux featuring the children's cuddly toys *Sylvanian Families*. The exhibit[6] was removed from the "Passion for Freedom" exhibition at the Mall galleries after police raised concerns about the "potentially inflammatory content" of the work (it was critical of the Caliphate). The Metropolitan Police, supposedly protectors of our freedoms, threatened the exhibitors – the potential victims – with a large bill for security if they continued to show Mimsy's little tableaux. Blackmailed, the gallery withdrew Mimsy's harmless little exhibit. (Three guesses as to why *Mimsy* is anonymous.) Jihadists 1; freedom, 0 in this match.

Fear suppressed the Mimsy exhibit. Fear is also the reason we are yet to see (say) *Kor Ani Get Your Gun*, or *All About Mohamed*, but *The Book of Mormon* runs and runs.

Fear is rarely so openly acknowledged. Instead, "sensitivity", the wish to hurt no one's feelings, is offered as the reason for banning, withdrawing or even condemning "offensive" material. In 1993 I wrote an *FN* editorial subtitled *'Sensitivity' and the Soft Option,* pointing out that when criticism is muted or censored, "sensitivity" is offered as the reason when, all too often, "what has in fact happened is that an act impelled by *fear* has been presented as the outcome of compassion... look back over press reports and count how many times decision makers have shown 'sensitivity' when there is, or is not, an implied threat of disruption. Our guess is they will find administrators much less sensitive when the protesting group poses no threat." And all this before 9/11, Isil, *Charlie Hebdo,* the Paris and Brussels attacks and the daily horrors on the evening news – threats not just annoying but mortal. Now political correctness can add

---

[6] http://www.theguardian.com/artanddesign/2015/sep/26/sylvanian-families-isis-freedom-of-expression-exhibition

"Islamophobia" to its other silencers: "racism", "sexism" and "homophobia". If academic leaders couldn't stand up then to a few angry students, what hope is there for us now when they may face real threats?

The problems with 'diversity' as a core aim of the university remain and may even have grown over the years. In 1988 the Academic Council at Duke voted on what came to be known as the Black Faculty Initiative. It proposed a numerical goal to increase the number of African-American faculty over a five-year span. The Initiative was up for renewal in 1993. The renewal was contentious. I wrote in a *FN* editorial: "The resolution...has never been freely discussed. It was passed under circumstances that were not conducive to calm deliberation. Public attention had been aroused; the Academic Council Meeting was packed with student activists; President Brodie announced his support before debate began; TV cameras were just outside the door. The atmosphere was intimidating." Even so, the critical vote was a tie – which was broken by the Chair of the Academic Council in favor of adoption. Affirmative action, as racial preferences are euphemistically termed, seems to be here to stay despite a couple of Supreme Court decisions that have weakened it at the margins.

Duke President, psychiatrist Keith Brodie was front and center on this issue. His aim, expressed in a letter to me was to achieve a "critical mass" of African American students at Duke. Academic qualifications were secondary. I pointed out that if the aim was to integrate minorities into Duke, rather than set the stage for self-segregation, "critical mass" was a really bad idea...

Now the drive for diversity has expanded well beyond African Americans. It extends even to the number of sexes. Two are apparently insufficient. The lines between male and female are to be eroded at the same time as their competitors are encouraged: to "L" and "G" are to be added "T", "B", "Q", and doubtless letters I've yet to encounter. Retired Keith B, his psychiatric training coming to the fore, has been a leader. Duke now has a Center for Sexual and Gender Diversity with several full-time staff. After reading through a web page entitled QUEERING DUKE HISTORY, which traces progress from "The Age of Arrest" in the 1960s to "The Age of Growth and Vitality" in the present, one can listen to a podcast of Keith musing about the history of the program, the involuntary nature of 'gender orientation' ("It's like left handedness"), and adding it to Duke's list of things we don't discriminate against.

Keith Brodie is a decent fellow; an excellent committee man, as I can attest having sat on a committee with him. But like almost all university presidents, he is much more concerned to calm waves than make them. If 'diversity' is what the loudest voices want, we're cool with that – especially if we agree with it anyway.

What is one to make of this "War on the Normal", this elevation of the margins?

> Turning and turning in the widening gyre
> The falcon cannot hear the falconer;
> Things fall apart; the centre cannot hold;
> Mere anarchy is loosed upon the world,
> The blood-dimmed tide is loosed, and everywhere
> The ceremony of innocence is drowned.
> The best lack all conviction, while the worst
> Are full of passionate intensity.

Not "best" and "worst" really, but "normal" and "abnormal" certainly. Otherwise, Yeats does seem to apply. Should one laugh or cry?

During the *FN* era, Duke 'diversity' administrator Leonard Beckum (quote: "I am the conscience of the university") said that "tolerance" is a bad word, because it implies that the thing tolerated is not respected. What's wanted is not toleration but celebration of things, practices, and people that were once ignored if not deplored. What's wanted is *status*; toleration is not enough. But where is the loving mother who *smiles* when her handsome son announces that he is gay? Just how desirable, never mind possible, is a complete inversion of the original hierarchy? If 'white privilege' is to be replaced by 'black/gay/transgender privilege' is the world a better place? Isn't the new hierarchy at least as odious as the old? Or should we be seeking a world of perfect equality, where Parkinson sufferers may aspire to become brain surgeons and dyslexics flourish as copy editors? The quest for perfection, for utopia, has led to some of the worst horrors of the twentieth Century from Stalin and Hitler to Mao and Pol Pot. Is that where we are headed?

A serious problem with 'diversity' as an institutional objective is that it fits the needs of the bureaucracy so much better than, say, excellence. *Excellence* as an aim has three problems. First, it's hard to measure. It involves effort, talent, curiosity, creativity and a host of other things that cannot be easily assessed. Second it inevitably involves *ranking*: some are more excellent than others. Worst of all, racially or ethnically identified groups will not be evenly arranged in such rankings. As the *Economist* commented at the time: "For a multi-ethnic society dedicated to the notion of equality, the idea that there might be a difference between the abilities and aptitudes of different racial groups remains the great taboo[7]." Assessing diversity, on the other hand, just involves ticking boxes.

---

[7]*Economist*, 9/12/92

The debate on political correctness at Duke reached a crisis point in late 1992. The London *Daily Mail* was then rather less focused on the anatomy, wardrobe malfunctions and sexual difficulties of celebrities than it seems to be now. It was no less sensational – but it did deal with a few serious issues – such as censorship. And it was influential. Nevertheless, I was surprised to receive from a faculty member a splashy two-pager, from the September 14, 1992 paper, that focused on Duke. The headline was "This conspiracy to rule our minds: They terrorize America, now Britain is their next target..." There were pictures of three U.S. academics: two small, each labeled *Victim*, and a large picture of Duke's own Stanley Fish, labeled *Persecutor*. Fish is a distinguished Milton scholar with very broad interests. I knew him quite well; I shared his distaste for Volvos, beloved of faculty (dull), and delight in Jaguars (sexy though unreliable). He himself owned a red (natch) 12-cylinder XJS. He had written a famously provocative article "There's no such thing as free speech, and it's a good thing, too." He argued that the author is irrelevant to the text and that all teaching is seduction (I wondered if the converse is also true: should I increase my course load?).

But Fish contains multitudes. Just a short time after the *Mail* incident *FN* published an article of his that began "The thesis I wish to pursue today will sound retrograde to many because it goes against the grain of much that has been said in recent years about literary and cultural studies. Specifically, I shall be questioning the possibility of transforming literary study so that it is more immediately engaged with the political issues that are today so urgent, issues of oppression, racism, terrorism, violence against women and homosexuals, cultural colonialism, and so on." Fair enough. And when Stanley left Duke, ascending the administrative ladder, for Chicago Circle, for Florida and finally the Cardozo Law School in NYC, his writings, in the *New York Times* and elsewhere, appeared increasingly moderate, sensible, even. His move from English to law is a natural one for someone who can so brilliantly argue any case.

But in the late 80s, Fish, then chairman, transformed the Duke English Department by hiring a slew of 'cutting edge' stars, such as Barbara Herrnstein Smith, queer theorists Eve Kosofsky Sedgwick and Michael Moon, and African-American-studies star Henry Louis "Skip" Gates Jr. Skip lived, briefly, in a large plantation-style house diagonally opposite us on Kent Street that had been used earlier as a set for the 1990 movie *The Handmaid's Tale*. (I can remember driving on a misty evening by the many parked trailers surrounding the set in 1989 with my parents who were visiting.)

Skip stayed for only two years, referring to Duke, rather unkindly, I thought, given the big house and other benefits, as a "plantation" (perhaps he was just referring to his house?). He left for Harvard where he later famously had an altercation with the Boston police. He was stopped for breaking in – to his own house! – in Cambridge in 2009. This incident led to a 'racial moment' with President Obama, after Obama called the Cambridge police "stupid"[8]. The President invited Gates, Vice-president Biden and two of the Cambridge officers to a tête-à-tête at the White House, confirming the anti-racist credentials of all concerned.

After Gates left Durham, the Kent Street house was bought by Michael Peterson a writer, Defense Department employee and, later, convicted murderer of one, and possibly two, wives[9]. Was there something in the water?

Perhaps it was all this activity by Stanley Fish that brought him to the *Mail's* attention. In any event I thought their rather sensational spread was both entertaining and informative: how do the Brits think of academic America? Of Duke? At one time we were internationally famous for J. B. Rhine and parapsychology – a reputation that was a constant embarrassment to the Psychology Department. Are we now to be known for political correctness? So I scanned in the two pages and printed the result in *FN*. I never heard from Stanley, but my guess is that he was amused. After all, he was the model for Brit novelist David Lodge's Morris Zapp, in the *Changing Places* trilogy. Fish/Zapp was used to publicity and showed no aversion to it.

But the spread outraged some of his colleagues. Cathy Davidson, then Professor of English and now Distinguished Professor and Director of the Futures Initiative at the Graduate Center of the City University of New York, condemned *FN* in the strongest terms. *FN* published her letter to the Chair of the Council, which concluded: "What does it mean that *Change* and U.S. *News and World Report* praise the English Department[10] 'built' by Stanley Fish at the same time that our own faculty newsletter engages in the most tasteless and malicious kind of colleague-bashing? I'm very glad there are no national rankings based on faculty newsletters. Ours is a disgrace."

Support came not just from Stanley's colleagues, President Brodie himself weighed in. In his state of the university address, he condemned the reprint, arguing that the *Newsletter* risked calling into question Duke's reputation for collegiality, continuing that the "level of personal attack is, to say the least, unbecoming, and serves no useful intellectual purpose." In a letter to the Chair of the Council, Brodie accused *FN* of "vilifying"

---

[8] The police handcuffed middle-aged Gates who walked with a cane. Maybe "stupid" is too kind: "nasty" might be better.

[9] Peterson appealed his conviction and the process continues as of September 2015.

[10] But see *Lingua Franca* February 1999, *The Department that Fell to Earth: The Deflation of Duke English.*

Professor Fish via "scurrilous republication". Wow! Brodie wondered whether Duke needed *FN* at all. I reprinted his letter, and my response, to the Council Chair, Richard Burton, of the Business School.

The fracas, and Brodie's hint of censorship, caught the attention of the local press. I was interviewed by the Raleigh *News and Observer*, with a picture, standing behind a table strewn with past *FN* issues. "Newsletter bedevils Duke" said the headlined (geddit? As the Brits would say[11]). Despite their usual bias, the press was against censorship and thus supported *FN*. *FN* survived, but my term came to an end in 1994. Keith Brodie's term also ended, in 1993. He was succeeded by Wellesley president Nannerl Keohane.

But the fun part of *FN* for me was not these momentous topics. I started some regular humorous and/or provocative columns that ran in almost every issue. For example, *Talking Points* included items like these, which do have some contemporary relevance:

Talking Points

Diversity: If they talk about diversity, they're gonna get it. If they talk about tolerance, they better be ready to have it.

(Harvard undergraduate Bridget Kerrigan, interviewed by the Boston *Globe* about the controversy caused by her hanging a Confederate battle flag from her dorm window 10/92)

The Norm: It was easy to list the guys I had slept with [in the last year]... but when I counted 24 I was like, gosh!... I'm out there to prove that it does happen to good, everyday people.

(Teenage AIDS victim, *Newsweek*, 8/3/92)

Another regular column reprinted striking examples of academic prose, inspired by Duke's own winner of the International Bad-Writing Award. The award was invented by Denis Dutton (1944-2010), professor of philosophy at the University of Canterbury in Christchurch, New Zealand and also a co-founder and co-editor of the excellent *Arts & Letters Daily* website. Here is a winning B-W example:

> The first prize goes to the distinguished [Duke] scholar Fredric Jameson, a man who on the evidence of his many admired books finds it difficult to write intelligibly and impossible to write well.

---

[11] The Blue Devil is Duke's mascot.

Whether this is because of the deep complexity of Professor Jameson's ideas or their patent absurdity is something readers must decide for themselves. Here, spotted for us by Dave Roden of Central Queensland University in Australia, is the very first sentence of Professor Jameson's book, *Signatures of the Visible* (Routledge, 1990, p. 1):

"The visual is *essentially* pornographic, which is to say that it has its end in rapt, mindless fascination; thinking about its attributes becomes an adjunct to that, if it is unwilling to betray its object; while the most austere films necessarily draw their energy from the attempt to repress their own excess (rather than from the more thankless effort to discipline the viewer)."

The appreciative Mr. Roden says it is "good of Jameson to let readers know so soon what they're up against." We cannot see what the second "that" in the sentence refers to. And imagine if that uncertain "it" were willing to betray its object? The reader may be baffled, but then any author who thinks visual experience is essentially pornographic suffers confusions no lessons in English composition are going to fix[12].

Emotionally inflated proclamations of the blindingly obvious are also irritating. They seem to have increased in recent years providing the "obvious" hits a fashionable spot. Alas there is no *FN* to mock them. Here is an example, selected as one of the writer's "most moving passages", by a recent *New York Times* reviewer: "Slavery is not an indefinable mass of flesh. It is a particular, specific enslaved woman, whose mind is active as your own; whose range of feeling is as vast as your own; who prefers the way the light falls in one particular spot in the woods. . . ." What might we learn from this? That slavery is bad? That slaves are people too? That some slaves are women? That all women have equally active minds and equal ranges of feeling? Everyone knows the first three and the last is obviously false (is a random slave, a random anyone, likely to have a mind "as active as" say Jane Austin or Queen Latifah? What nonsense!) What then is "moving", much less informative about this passage?

Here are some more literary jewels, from *FN* in the early 90s.

---

[12]http://denisdutton.com/bad_writing.htm

Communication Skills

Gems of scholarly prose from around the world.

This month: bipolar themes from the New Feminist Scholarship.

'Yes' is not different from 'no', therefore men dress up as women?

> The current popularity of cross-dressing as a theme in art and criticism represents, I think, an untheorized recognition of the necessary critique of binary thinking, whether particularized as male and female, black and white, yes and no, Republican and Democrat, self and other, or in any other way.

> Marjorie Garber in *Vested Interests*
> [Prof. Garber is currently the William R. Kenan, Jr. Professor of English and of Visual and Environmental Studies at Harvard.]

Pigeonhole Problem

> The relative difficulty with which oral sex, as opposed to anal, can be schematized in the bipolar terms of active/passive or analogically male/female, would also seem congruent with the process by which the trope of gender inversion was giving way to the homo-trope of gender sameness.

> Duke Distinguished Professor Eve Kosovsky Sedgwick in *Epistemology of the Closet*

> Sexuality becomes a property of the individual the more the lifespan becomes internally referential and the more self identity is grasped as a reflexively organised endeavour.

> Antony Giddens, in *The Transformation of Intimacy: Sexuality, Love and Eroticism in ModernSocieties*. [Baron Giddens of Southgate, a Labour Peer, is the author of more than 30 books in 29 or so languages. He is one of Britain's most distinguished sociologists.]

As you can imagine, satirizing academic prose, especially the prose of one's colleagues was not popular. But it was not fatal. In retrospect, my colleagues tolerated *FN* and treated its editor remarkably well. And maybe Duke's academic prose improved a bit? All that seems to remain is my reputation as a mildly controversial figure, the details of which are lost in the mists of antiquity.

My favorite humor column was *Rumor and Innuendo,* tag line: "Heard on the Quad by 'Possum". Here are a few samples, October 1992 and February 1993 (the influence of *Private Eye* is undeniable). First, is a prescient spoof, featuring financier Michael Milken, inventor, or at least exploiter, of the junk bond. Milken was imprisoned for racketeering and securities fraud in 1990, although, as in so many prosecutions of visible Wall Streeters, the line between fraud and legal ingenuity is often hazy. After Milken was released from prison, in 1993, he created a foundation for cancer research. This item features a quote from Larry Summers, then Chairman of the World Bank, which got less criticism than his later comments on women in science, mentioned in Chapter 3.

Rumor and Innuendo

The visible hand

We just got an enthusiastic note from our esteemed colleague, Professor Adam Schmidt, holder of the Ivan Boesky[13] Chair of Welfare Economics in the newly founded Milken School of Business in Beverly Hills, CA. Schmidt writes:

Hail Possum! All of us here at MSB are excited about the recently released comments of Lawrence Summers, chief economist of the World Bank. Writing to fellow money-persons in the bank, Summers raised the following question: "...shouldn't the World Bank be encouraging more migration of dirty industries to the LDCs [less-developed countries]? ... [From the point of view of earnings foregone] a given amount of health-impairing pollution should be done in the country with... the lowest wages. I think the economic logic behind dumping a load of toxic waste in the lowest-wage country is impeccable..."

We all agree, of course, and would like to extend the argument to the PDCs (partially developed countries) – the U.S., for example. In fact, we can go Summers one better: why not put toxic dumps in high-welfare areas, thus eliminating individuals who are not just low-wage but who actually cost money? Wow, what a saving! Hospitals,

---

[13] Ivan Boesky was a Wall Street trader who testified against Milken and was indicted for insider trading in 1987. He was the "greed is good" model for Gordon Gekko in the movie *Wall Street.*

particularly those with a high proportion of terminally ill individuals, are another "natural" – especially as they generate a lot of the stuff themselves. Even as we speak, our Toxic Waste Investigation Team (TWIT), is seeking still more cost-wise applications of Dr. Summers' exciting new insight.

Yours in solvency, A. Schmidt.

And from February 1994

The eyes have it.

Some male faculty members in the Religion Department got an anonymous message in their mailboxes a few weeks ago. The note was a list of do's and don'ts on male-female behavior. One memorable "do": look women in the eyes; looking anywhere else is evidently *verboten*. [This must be why so many women wear low-cut tops – so that men will look elsewhere?]

Diversity – the Next Step

The underrepresentation of oppressed groups – blacks, women, Republicans – in the academy is a continuing disgrace. A few hesitant steps, such as Duke's new Black Faculty Resolution, are being taken to remedy at least some historical inequities. Concerned administrators are sensitive to many relevant issues. But campus groups such as the middle-of-the-road *Spectrum* and activist newcomer *Broadband* are pointing to a question that seems to have been completely neglected, namely *course enrollments*. This is a university-wide problem. Faculty observers have already noted enormous disparities in enrollments. The racial and gender composition of classes varies in an uncontrolled and clearly discriminatory fashion across disciplines. Women, for example, are vastly underrepresented in engineering courses. Men are not enrolled in proportional numbers for courses in Women's Studies and child psychology. Whites are underrepresented in African-American Studies, blacks in Caucasian Studies [some mistake here, surely – Ed.].

Show us a crowd, and we will follow!

Duke, always in the vanguard on numerical issues, has formed a committee to look into this problem, the Course Representation and

Balance Committee (CRAB) – a successor to the much-criticized Course Representation and Proportion Committee which dissolved in mutual recriminations last Fall. CRAB's first action was of course to recommend the appointment of a Vice-Provost for Balance "to coordinate balance efforts throughout the University." When serious debate began, it soon reached an impasse. Two opposing factions developed. The traditionalists, led by Professor of Xerox Studies Justin Tyme, are committed to full proportional representation: every course must achieve an exact match between enrollment proportions and proportions in the University as a whole. The progressives, led by Associate Professor of Critical Practice Aspidistra Virago-Dworkin, favor a more activist approach: course enrollments should encourage "proactive remediation," a concept advanced in a recent *Nation* article. The method involves a complicated mathematical formula, but the basic idea is that students should be allocated to courses at the bottom of their preference lists. "It's the only way to correct false consciousness and past discrimination" said Virago-Dworkin, in an interview.

Committee Chair, Vice-Provost Alvin "Stroker" Neutral, could give no date for the committee's final report.

Heard in the Quad by *Possum*

I will end on a historical note. I have always enjoyed reading the wonderful diary of Samuel Pepys (1633-1703). Pepys wrote the diary in a sort of shorthand from 1660 ceasing in 1669, when he fancied that his sight was failing (it was not). The diary was rediscovered and translated in the 19th century. It covers an exciting period in English history, including the plague and the great fire of London. I used to read bits to my wife at night, in the days when I was still awake at a late hour.

In any event, in the early 1990s, a rumor circulated at Duke about sexy carryings on at the pinnacle of the Medical Center. Apparently (the rumor ran) a senior person at DUMC had been caught by a cleaning woman in a compromising position with a lady assistant on a table in his office. Nothing official was ever heard and the cleaning lady was not to be found. The rumor prompted the following *Possum* column in which I reprinted a portion from Pepys' little-known *New Journal*, written after he realized that he had not lost his sight, some years after the original:

London Journal: The New Diary of Samuel Pepys, Gent.
IOVRNALL. Vol. XX
January 4. 1671/72—Lords-day.

Up betimes and by coach to the College of Chirurgie in White-hall, established nigh these forty years by my Lord Duke for the care of the sick (and enrichment of the Guild of Chirurgeons). Much activity though it be a Day of Reste, and to my great amusement all agitated by tales of the frolics of the Physician-in-Ordinary to His Majestie, Sir Aubrey Snoot. Sir Aubrey hath been observed <u>in flagrante</u> by a sweeping wench and Sir W. Batten gave us an account of his prowess in the Arts of Venus with a Lady of the Court. Sir Aubrey's feat of gallantry excited much admiration for the dis-ease of his field of battle, <u>non campus sed mesa</u>.

5. At my office, where not much work done to-day for discussion of Chirurgical matters. Mr. Sheply spoke upon the matter of the College and the ars amatoria, calling us to remember Dr Brindsley, a physician and Minister of the Church (tho' a dissenter and low man say some). Many were the wenches and even ladies of quality comforted by Dr Brindsley in his apartments in the College. Mr. Shepley hath entered into some study of the matter and discoursed upon the foetid and perfumed air of the 'Spital and its effect upon the amatory propensities of old physicians. His reason was strong and all agreed that no blame attached to these gentlemen for their amatory enthusiasms. But Dr. B. says they cannot claim especial Powers, for their spirits are excited by a sort of phisic.

Early to bed, but some difficulty sleeping.

Some years later I had a correspondence with the writer Tom Wolfe. The topics were neuroscience, linguistics and the culture wars. Wolfe's irritation at obscure or jargon-filled writing is possibly even greater than mine. Witness his comments on art criticism that I quoted in Chapter 3. I sent him a few **Communication Skills** examples, which he seemed to appreciate.

I sent this extract to Mr. Wolfe who was generous in his appraisal, saying my "rendition of Pepys is absolutely brilliant!... it was a delight to read." I was flattered, but of course all the credit goes to Sam P.

The piece was never actually published in 'Possum' because forbidden by the authorities[14].

---

[14] My wife.

Epilogue

Why *The Englishman*? First, because my friends and colleagues in America seem always to think of me in that way. But more importantly, because I cannot ignore the intellectual lodestars of my youth. The first twenty years are a sort of 'critical period' I think, when one imprints – like a duckling on its mother – on the ideas and people that are landmarks for later intellectual development. I am much more likely to think of Francis Bacon than Montaigne, of Bertrand Russell than of William James, of Peter Simple than H. L. Mencken and of Sam Johnson rather than Ben Franklin. As my education advanced, my intellectual horizons broadened. But those early landmarks will always remain.

The theme of my professional life has been trying to understand how animals learn. One thing I *have* learned is that science, especially psychology, is mostly the exploration of dead ends. The past one hundred years is a history of *culs de sac*, each born with a flourish and then slowly fading away. Part of the problem is that anything touching on human experience immediately encounters ready-made accounts, common-language labels, folk-psychology explanations that take many years to see beyond. Historically, the first step, introspectionist psychology, was to take common sense at face value and try to make the labels more precise, more scientific. Another tack was to avoid introspection completely by assimilating everything to the model of the reflex. Yet a third was B. F. Skinner's complete abandonment of any notion of mental or physiological processing: a 'science of behavior' that embraced stimuli and responses and nothing else. All these have failed. Only now, after very many years, do we seem to be settling on a Darwinian structure: learning as a process of selection from a repertoire, partly expressed, but mostly covert – processes of behavioral variation analogous to the genetic variation which allows natural selection to work. Is this view perhaps also a dead end? I don't think so, but in science nothing stays fixed for long.

This book covers only the first fifty or so years of my life. In later years, I became interested in social as well as biological topics: economics and policy. I wrote articles for mainstream publications on issues such as personal responsibility and free will, finance and reinforcement psychology, and traffic control and perceptual psychology. *Distracting Miss Daisy,* as the editors felicitously titled it, got the most attention but has had minimal effect on the long-established inefficiencies not to say hazards of the American system of traffic control.

Irritated by the false information circulated about the evils of smoking – which is risky, but far from lethal – I wrote a little book, *Unlucky Strike,* reviewing the science and the ethical issues. The science is not conclusive even about secondhand smoke and the economics says that smokers don't cost the rest of society. Conclusion: smoking is a private not a public health issue, so the government should stick to (accurate!) information but otherwise – geroff! We got to know artist David Hockney, who at the time lived not far from York and very kindly provided a few illustrations of ashtrays and other historical curiosities.

I wrote a book on the regulation of financial markets – I even lectured about it on Wall Street. Most of my necessarily naïve (I'm not a professional economist) diagnoses of the 2008+ financial crisis seem nevertheless to have been vindicated. The crisis was the product of a system where benefits to financial players were immediate, but costs were delayed and fell on others. This is what I called *The Malign Hand.*

How might the problem be fixed? It seems obvious that regulation should control the payoff schedule of individual agents. It should ensure that financial players pay the costs as well as reap the benefits of their own behavior. Instead we have what I called "regulation by scrutiny", where armies of underpaid bureaucrats try to figure out whether financial agents – smarter, more highly paid and certainly more motivated – are doing things that might be 'systemically hazardous'. What could possibly go wrong!

Trying to recreate my own past life has been a curious experience. Retrospection has never been my strong suit. My memory for personal details is not great. Colleagues, fellow graduate students at Harvard, say to me "John do you remember when you..." Usually, I don't. My sister Judy has a much better 'personal memory' than I, although even she took many years to finish a memoir of her teenage years in Africa. Yet I have been able to recreate enough of an odd life to be I hope of general interest. I have lived through great events, such as the London blitz and the dissolution of the British Empire, globalization and its costs and benefits, not to mention America's mini-wars and financial crises.

My education has been a series of happy accidents. I got a good grounding in 'hard' science and math in my grammar school, although my record there was not impressive. But I was blocked from my real love, biology, by the limited resources of post-war England and the rigidities of the English educational system. Nevertheless, all was not lost. I wound up, via a devious route, in psychology, a field that allows great freedom. I was able to become involved in biology, ecology, economics, and philosophy as well as more conventional psychology. A specialization in biology might not have allowed me to stray so far.

I entered the world of academe at a lucky time. Demand for new PhDs was high, research was largely unregulated, costs were modest and grants could be got with an amount of effort that did not subvert their purpose. Then, the application was short and success likely; now, getting a grant takes more time than actually doing the research and success is far from guaranteed. So I shut my lab and quit trying just a few years ago and entered on a happy retirement – with time to write a memoir.

Acknowledgements

Many friends and relatives shored up my imperfections of memory, syntax, spelling and clarity. I thank especially my wife Lucinda, sister Judy Rawlinson and my children Nick and Jessica. Thanks also to Phoebe, Patricia, Katharina, DB, Jennifer Higa, Mike Davis, Alliston Reid, Ken Steele, Alex Kacelnik, John Malone, Lino Bueno, Silvano Zanutto and Emily Evans of the De Cordova Museum in Lincoln, MA. Christopher Woodhead at the University of Buckingham also helped with my earlier book *Unlucky Strike.* He has made the final copy-editing and production of both books a pleasure rather than the trial it so often is. And finally, apologies to those I have forgotten!

Cover: England, Burma, India – and psychobiology. Top is a Victorian-era watercolor I bought in 1950 or so. Bottom is a picture by a Burmese artist my mother bought in Rangoon in 1936. Inset is a pigeon perching on a wall in the Amber Fort, Jaipur, India some years later.